WRITE & WRONG

SUCCESSFUL WRITING STARTS WITH KNOWING THE DIFFERENCE BETWEEN

STUDENT WORKBOOK

WRITING WITHIN CRIMINAL JUSTICE

CAROLINE W. FERREE
School of Criminal Justice
University of Baltimore

HEATHER L. PFEIFER
School of Criminal Justice
University of Baltimore

JONES & BARTLETT
LEARNING

World Headquarters
Jones & Bartlett Learning
5 Wall Street
Burlington, MA 01803
978-443-5000
info@jblearning.com
www.jblearning.com

Jones & Bartlett Learning books and products are available through most bookstores and online booksellers. To contact Jones & Bartlett Learning directly, call 800-832-0034, fax 978-443-8000, or visit our website, www.jblearning.com.

Substantial discounts on bulk quantities of Jones & Bartlett Learning publications are available to corporations, professional associations, and other qualified organizations. For details and specific discount information, contact the special sales department at Jones & Bartlett Learning via the above contact information or send an email to specialsales@jblearning.com.

Production Credits
Publisher: Cathleen Sether
Acquisitions Editor: Sean Connelly
Editorial Assistant: Caitlin Murphy
Production Manager: Tracey McCrea
Production Assistant: Eileen Worthley
Marketing Manager: Lindsay White
Manufacturing and Inventory Control Supervisor: Amy Bacus
Composition: Cenveo Publisher Services
Cover Design: John Pfeifer/ Kristin E. Parker
Printing and Binding: Edwards Brothers Malloy
Cover Printing: Edwards Brothers Malloy

6048

Printed in the United States of America
16 15 14   10 9 8 7 6 5

Dedication

In memory of my mother, who always believed in me.
— Caroline Ferree

To my husband, John, and my daughters, Emily and Addy, for their never-ending patience
and encouragement.
— Heather Pfeifer

Contents

Preface

This workbook is designed specifically to help criminal justice students improve their research and writing skills. It can be used as a class text and as a reference guide for students to use outside of class. By using this workbook, students will learn how to find academic sources through library research, how to organize research material and use it to write a paper that follows *APA Publication Manual* (6th edition) rules, how to create a reference list for a paper that follows APA style, and how to create a résumé.

To help students practice the different research and writing skills that are covered in this workbook, most of the units include handouts that they can refer to when working on a paper, as well as exercises that will help them to practice the skills that were introduced in that unit. In addition, each unit in the workbook includes fill-in-the-blank examples for students to answer. Finally, at the end of each unit there are blank pages on which the students can write notes.

For some students, this workbook will serve as the main text for a research and writing course that they must complete within their specific degree program. As such, there are handouts and class exercises included at the end of each unit. For other students, this workbook will be used as an ancillary to the main text. In this case, the instructor will incorporate one or more of the units from this workbook into his or her regular course content to teach specific research and writing skills (e.g., plagiarism, APA rules of citation). Because the instructor might teach only specific units, he or she will tell the students which handouts and which units to use.

An *Instructor's Manual* (978-1-4496-2682-2) is available for adopting institutions and contains lectures corresponding to the chapters of the ***Student Workbook***. Every lecture includes detailed lesson plans, Notes to the Instructor, PowerPoint presentations, in-class exercises with answers, and reference guides. Also contained in this in-depth teaching tool are a sample syllabus, grading rubrics, homework assignments with answers, and a midterm exam with answers.

Acknowledgments

I would like to thank Kim Wiklund for her endless patience, continuing support, and unwavering sense of humor throughout the course of the writing of this book. Her contributions were invaluable. I would also like to thank my coauthor, Heather, for asking me to help her grade student papers over a decade ago, which set into motion the creation of a research and writing class for criminal justice students and, ultimately, this book.

— Caroline Ferree

I would like to thank my mentor, David Barlow, for helping me to discover my love of teaching and encouraging me to write this book. I would also like to thank my coauthor, Caroline, for remaining in the trenches with me these past several years and helping to bring this project to fruition.

— Heather Pfeifer

We would also like to thank the following individuals who reviewed the manuscript:

David Burlingame, Grand Valley State University

Helen Eigenberg, University of Tennessee at Chattanooga

Cynthia Koller, Florida Southern College

Billy Long, Ferrum College

Sharon RedHawk Love, University of Tennessee at Chattanooga

Michael J. Montgomery, Tennessee State University

Daniel Philips, Lindsey Wilson College

Melinda D. Schlager, Texas A&M University–Commerce

Introduction

UNIT SUMMARY

Learning Objectives

At the end of this unit, students will be able to do the following:

- State the general rules about academic writing.
- Identify academic sources and the databases in which to find them.
- Identify the difference between informative and persuasive papers.
- Narrow the scope of a too-broad topic and create a thesis statement for the new topic.
- Write a problem statement for a paper.

Introduction

Using this workbook, you will learn how to find academic sources through library research, organize the materials, use them to write a paper in accordance with the *Publication Manual of the American Psychological Association* (APA), *Sixth Edition*, create an APA-style reference list for the paper in accordance with the *Publication Manual* (6th edition), and write a résumé.

There are four handouts included at the end of this unit. The first handout, "Sample Problem Statement," relates directly to this unit, and you will work with it at the end of this unit.

Handouts #2 through #4, listed below, are homework assignments that will be referred to throughout the workbook. However, these assignments might not be used by every instructor. Accordingly, if your instructor does decide to use them, he or she will let you know when you need to refer to them.

Introduction, Handout #2: "Take-Home Assignments"

Introduction, Handout #3: "Writing Assignments"

Introduction, Handout #4: "Job Profile Assignment"

Writing a Paper: An Overview

As a criminal justice student, you will be asked to write several papers. For this class, you must write a research paper. Students are often confused about what this means. It is important that you understand what a research paper is so that you will know what you must write for this class.

The Research Paper: Primary Versus Secondary Research

For a *research paper*, you will summarize and critique research that has been conducted by other people. In other words, you will formulate a thesis statement (which you will learn how to do later in this unit) and do library research to find published literature on your topic. You will then read and summarize that information. This is called *secondary* research. It is different from *primary* research, in which researchers collect and analyze data, make findings about it, and draw conclusions from it. Typically, scholars in the field write papers using primary research.

> **Note!**   For the purposes of this workbook, the terms "research" and "library research" mean the same thing.

The Importance of Good Content and Good Presentation

An effective paper must be strong in both content and presentation. To understand what this means, imagine the two scenarios in the following example:

Example:

Scenario #1: *You go to a wedding where there is a beautiful cake. It has fluffy, white frosting with lots of colorful flowers. It is perfectly shaped and looks spectacular. However, when you are given a piece to eat, it is dry and has no flavor. You are disappointed, because it looked so good that you thought it would taste equally great.*

Scenario #2: *A little girl wants to make her mother a birthday cake. With the help of her father, she follows a recipe for a chocolate cake. She insists on frosting and decorating the cake herself. When she is finished, it is lopsided and the frosting is uneven. But it tastes delicious.*

How do each of these scenarios compare with writing a paper that has both good content and good presentation?

Scenario #1:

_____

_____

Scenario #2:

_____

_____

The bottom line is that neither of these papers is going to earn a good grade, because you must have both good content and good presentation when you write your paper!

By using this workbook, you will learn how to construct a paper that is strong in both content and presentation. You also will learn how to do library research to find the information you need to write your paper and to present it in a manner that is well organized, well written, and consistent with the citation standards of the *Publication Manual* (6th edition).

Academic Writing: General Rules

Academic writing is more formal than other writing styles. The mechanics of writing will be covered later in this workbook; for now, here are a few general rules:

- Do not use contractions.
- Do not use slang.

UNIT
2

UNIT
3

UNIT
4

UNIT
5

UNIT
6

UNIT
7

UNIT
8

UNIT
9

Example:

Write two examples of slang you should not use and the correct words you should use in their place.

_____

_____

- Do not use flowery language; state things simply and clearly.

Example:

Write an example of flowery language and the correct (simple) way to say it.

_____

_____

- Present all of the information in a neutral and objective tone; you must not include your personal opinions.

Example:

Write five pronouns you should avoid using.

_____

- Include only information that comes from academic sources, not from your own knowledge.
- Cite all of your sources in your paper in APA style, consistent with the *Publication Manual* (6th edition).
- Create a reference list, prepared in APA style, consistent with the *Publication Manual* (6th edition). It must include all of the sources you cite in your paper.

Academic Sources

When you do your library research, you must obtain your information only from academic sources. An academic source is also known as a "scholarly source." It is empirically based; that is, it is grounded in research and is not simply someone's personal opinion. Moreover, an academic source is one that has been "peer reviewed" by experts in the field for its accuracy and quality. Some examples of academic sources are:

- *Peer-reviewed journal articles.* These are articles that have been "cleared" by scholars in the field before publication. In other words, experts in the field have reviewed them and recommended them for publication.

Example:

List two peer-reviewed journals:

_____

_____

- *Scholarly books.* These are good resources because they often present a wide range of information on a topic written by experts in the field. A scholarly book can be a summary of multiple authors' (or a single author's) own research, or it can be a volume of related essays—similar to an anthology—written by several authors and compiled into a book by an editor.

Examples:

Paternoster, R., Brame, R., & Bacon, S. (2008). *The death penalty: America's experience with capital punishment.* New York: Oxford University Press.

Wilson, J. Q., & Petersilia, J. (Eds.). (2011). *Crime and public policy.* New York: Oxford University Press.

To determine if a book is scholarly, read the preface or introduction to see if it lists the authors' credentials (e.g., PhD) or affiliations with educational institutions or government agencies. If the authors' credentials or affiliations are stated, it is probably a scholarly book.

- *Research reports published by government agencies.* Two good sources for research reports are the National Institute of Justice (NIJ), which is the research arm of the U.S. Department of Justice, and the Bureau of Justice Statistics (BJS). The BJS will likely have any crime-related statistic you require. Other good sources for research reports are state government agencies that have a publishing or research department, as well as independent social research agencies.

Example:

Write the names of two government research agencies, other than the NIJ and BJS, that publish criminal justice-related research.

_____

_____

- *Law review articles.* Published by law schools, law review articles are a good source if your topic is law related. However, their scope is typically limited because their focus is exclusively on the law.
- *White papers.* A white paper presents an agency's or organization's social or political position on a particular issue. Its purpose is to educate the public on how and why the agency views the issue the way it does. White papers occasionally make recommendations on how an agency thinks an issue should be handled.

Nonacademic Sources

Nonacademic sources do not require the same in-depth review process that academic sources do. Accordingly, you should *not* use any of the following to write your papers:

- magazines (e.g., *Newsweek, Time, U.S. News & World Report*);
- newspapers (e.g., *New York Times, Washington Post*);
- encyclopedias or Wikipedia;
- textbooks; or
- trade journals (e.g., *Police Chief*).

UNIT 2

UNIT 3

UNIT 4

UNIT 5

UNIT 6

UNIT 7

UNIT 8

UNIT 9

There are two exceptions to this list:

1. You can use encyclopedias specific to criminal justice or criminology, such as the *Encyclopedia of Crime & Justice*, the *Encyclopedia of Criminology*, and the *Encyclopedia of Crime and Punishment*. These academic resources provide overviews of specific topics that relate to crime and the criminal justice system, and using them may help you to better understand your topic.

2. For the paper you will write in this class, you may use a recent case profiled in the news to introduce your topic. However, because that information will come from a nonacademic source, you cannot and should not include it as part of your library research. Further, it will not count toward the required minimum number of academic sources you must use when you write your paper.

Identifying Academic (Scholarly) Sources

Sometimes you will not know if a source is academic (scholarly). In such cases, there are certain things you can check to help you determine whether it is a scholarly article.

- *Names of authors.* Scholarly sources have named authors. If an article is attributed to "Anonymous," it is not scholarly.
- *Length.* A scholarly article is typically 5–30 pages long. If the article is very short (e.g., one page), it is not scholarly.
- *Pictures.* Scholarly articles typically do not have pictures.
- *Reference list.* Scholarly articles are research articles, and, as such, the authors must include a reference list of their sources. If an article does not have a reference list, it is not scholarly.
- *Biographies.* Scholarly journals often include author biographies. Nonacademic journals do not. Note, however, that you should not rely solely on this factor to determine whether the source is scholarly, because not all scholarly articles include biographies.
- *Credentials (e.g., JD, PhD) after the authors' names.* As with biographies, scholarly journals often list author credentials; however, they do not *always* do so. Again, you should not rely solely on this factor to determine whether a source is scholarly.

Suggested Databases for Library Research

Students often use databases such as Wikipedia and search engines such as Google™ or Yahoo!® to conduct their library research. Do not do this! Most of the information contained in those types of databases and general search engines is not academic. Conduct your library research in criminal justice-related databases and websites.

Example:

List four criminal justice-related databases and websites.

_____

_____

_____

_____

UNIT 1
UNIT 2
UNIT 3
UNIT 4
UNIT 5
UNIT 6
UNIT 7
UNIT 8
UNIT 9

Note! Although these databases and websites are good to use for criminal justice research, do not limit yourself to these sites. Many criminal justice issues are multidisciplinary, and you can find criminal justice-related research in the works of other disciplines such as psychology, sociology, and education. Moreover, different disciplines approach the topics from different angles and, therefore, may provide you with valuable information. If you do use databases from other disciplines, remember to use your checklist to determine whether the articles you have found are scholarly.

Rules to Follow for Selecting Sources

As you do your library research, there are several ways to ensure that you find the best sources for your paper.

- *Read abstracts.* Abstracts give you a summary of the article or report, which will allow you to determine whether it is on point with your research question or thesis statement.
- *Focus on current research.* Limit your research to sources published within the past 10 years.
- *Focus on current statistics.* When you write your paper, you should present the most recent statistics available that are related to your topic.
- *Keep it in the United States.* Unless you are writing a paper on an international topic, you should use research that has been conducted only in the United States.

Two Types of Papers

There are two styles of academic papers: the informative paper and the persuasive paper. The purpose of both types is to educate the reader.

The Informative Paper

An informative paper enlightens a reader by presenting a summary of the literature relating to a topic. Your informative paper should educate the reader by answering the following questions: Who?, What?, Where?, When?, and Why? If the topic addresses a particular policy or issue, you should remain objective and neutral by presenting all sides.

The Persuasive Paper

When writing a persuasive paper, you will choose a position on a debatable issue relating to the topic and will argue that position by presenting empirical studies that support your argument. In general, an empirical study is one in which researchers have analyzed data from which they have made findings and drawn conclusions. Although there may be strong arguments that are contrary to your chosen position, you should acknowledge them and then present your research in a manner that will refute those arguments.

Note! When you write a persuasive paper, take a position but do not include your personal opinion. The word "I" should not appear in your paper.

The Term Paper for This Class

In this class, you will write a term paper about a program or policy chosen from the list of approved topics in Introduction, Handout #3: "Writing Assignments." In the paper, you must educate your reader about your program or policy by discussing the elements that are listed in the assignment. You must also discuss three empirical studies that have evaluated the effectiveness of the program or policy.

As explained in the handout, the first part of your paper will educate your reader by including the following information about your policy or program:

- a description of the policy or program,
- the scope of the problem it is trying to address,
- its history,
- its purpose or goals,
- the population targeted, and
- its activities or elements.

When you discuss the history, you must include information about its origins, as well as the context in which the program or policy was created.

 Hint!   When writing your paper, you should discuss each element in the order in which it is listed in the handout. Doing so will ensure that your paper is well organized.

After you have educated the reader about the policy or program, you will summarize three empirical studies that have examined its effectiveness. Again, Introduction Handout #3, "Writing Assignments," lists the information you must include in the discussion. Specifically, for each study, you must summarize the research question, identify the population examined, and discuss how the data were obtained, as well as the findings and any significant limitations. When you write this part of the paper, you must discuss each study separately; do not "lump" them all together into one paragraph.

You will write and hand in two drafts of this paper, which will be graded separately. The first draft will be graded and will contain feedback on its content, clarity, documentation (e.g., citations), and organization. You will incorporate those suggestions to edit your second draft, the final paper.

Narrowing the Scope of a Topic

In some classes, you may be asked to write a paper on a very broad topic (e.g., "female offenders"). If you were to attempt to research that topic, you would get hundreds of hits (search results)—an overwhelming amount of information. Therefore, you must narrow the scope of the topic before you start your library research. To do this, follow these steps:

1. *Brainstorm!* On a piece of paper (or in a Word document), list all of the answers you can think of to the following questions:

   a. *Who?* In other words, what types of females are there (e.g., age, race, ethnicity, role)?

Example:

Write five examples of different types of females.

_____

_____

_____

_____

_____

b. *What?* In other words, what types of offenders are there (e.g., drug offenders, violent offenders, sex offenders)?

> **Example:**
>
> Write three examples of different types of offenders.
>
> _____
>
> _____
>
> _____

2. Combine some of the elements from both lists to create a type of female offender.

> **Example:**
>
> Write examples of three types of female offenders.
>
> _____
>
> _____
>
> _____

3. Once you have a specific type of female offender in mind, ask yourself what you are interested in learning about this population. For example, Why might they engage in that behavior? or What is the community or criminal justice system doing to prevent or control the behavior? This will help you begin to formulate your research question or thesis statement.

> **Example:**
>
> You are interested in adolescent female gangs. One issue you could explore would be the risk factors that have been identified for girls who join gangs.

Note! When you choose a topic for a paper, choose something that interests you. It is easier to write a paper about something you care about than to write a paper about something that bores you.

Writing a Thesis Statement

A thesis statement is a one-sentence statement that tells your reader what your paper will be about. You must have a thesis statement when you write a persuasive paper. Thus, you will have to write a thesis statement for your term paper.

UNIT 1
UNIT 2
UNIT 3
UNIT 4
UNIT 5
UNIT 6
UNIT 7
UNIT 8
UNIT 9

How to Write a Thesis Statement

When you write a thesis statement, you should follow these rules:

- It must be specific and narrow.
- It cannot be a question.
- It must be a debatable issue for which research has been conducted. It cannot be a statement of fact or an opinion.
- You must pick a side of the debatable issue; do not be ambiguous.

Once you have narrowed the scope of your topic and determined what you are interested in learning about that particular type of offender, you can write a thesis statement. However, before you write a thesis statement, you should do some preliminary research about your topic so that you become familiar with it. This will make it easier to write your thesis statement.

When you do your preliminary research, you will look for the answers to particular questions.

Example:

Assume you have narrowed the scope of your topic to "juvenile female gang members." You would research the following questions:

- *Why?* Why do those offenders engage in the particular activity? For this example, you would look for answers to the question, why do juvenile females join gangs? In other words, what risk factors are associated with juvenile females joining gangs (i.e., drug use, childhood abuse, peer pressure)?
- *How?* How can the behavior be stopped? For this example, you would look for programs, policies, or intervention strategies that have been established to help young female gang members leave the gangs (i.e., drug treatment programs for youth, job training programs).
- *What?* What changes do we expect to see? For this example, you would look for discussions about the behavioral changes of young female gang members who participate in a drug treatment program or who are affected by a policy.

Once you have considered each question, you can put your findings together into a thesis statement.

Example:

Write a thesis statement for a paper about juvenile female gang members.

_____

_____

Note! After you start your library research, you may discover that there is too much information about your topic or that there is not enough. In either instance, you should edit your thesis statement. If there is too much information, you should try narrowing your thesis statement further. If there is too little information, you may want to broaden your thesis statement.

Writing a Problem Statement

Sometimes, instructors will ask students to write a problem statement for a paper. A problem statement is a paragraph that describes a problem you believe needs to be addressed and sets forth a solution for that problem. When you write a problem statement, you must clearly articulate for the reader what the problem is and must discuss why it is important that it be addressed. Before you write your problem statement, you must first think of a problem that needs to be addressed. Examples of problems you could address include prison overcrowding, the disproportionate representation of minorities in correctional populations, and the increasing number of juveniles waived to the adult system. When you select a problem, remember that it must be one that can be resolved!

Once you have determined the problem you will address, you can begin to draft your problem statement. There are three parts to a problem statement: (1) the vision statement, (2) the issue statement, and (3) the solution statement.

The first part of the problem statement—your vision statement—is the opening sentence of your paragraph. It describes what the ideal scenario would be if your problem were remedied.

> **Example:**
>
> In your problem statement, you will address the problems that sexual assault victims may encounter in the criminal justice system after reporting an assault. Specifically, they are often treated poorly by the criminal justice system. Your vision of an ideal scenario for these victims might be that they would feel safe and comfortable enough to report the assault, and that professionals would be available to help them navigate the criminal justice system. Accordingly, the first sentence of your problem statement (the vision statement) might read:
>
> *"When victims of sexual assault seek assistance from the criminal justice system, they expect that the professionals they encounter will treat them with compassion and respect and will help them navigate the various processes in the system."*

The second part of your problem statement is the issue statement. This will be longer than your vision statement and will consist of several sentences that describe the problem in its current state. Thus, for the above example, you would describe what the situation is really like for sexual assault victims who report their crimes.

Before you write your issue statement, you should spend some time thinking about the answers to the following questions:

1. Who does the problem affect (e.g., specific people, communities, agencies)?

   In the example, it affects sexual assault victims.

2. What does the problem affect (e.g., what is its impact)?

   In the example, it affects the victims' willingness to report assaults.

3. When does the problem occur (e.g., time, processes)?

   In the example, it occurs when the victims report assaults and as they move through the legal process. Specifically, victims could be asked to repeat their stories several times to different people. Also, they might not be told what to expect from the system, might not be kept apprised of how their case is progressing, and might not be told why certain decisions are made.

4. Where is it a problem (e.g., only in certain places, instances)?

   In the example, it is a problem in hospitals, other medical facilities, police stations, and courts.

5. Why is it a problem (e.g., What are the consequences, and why is it important we fix it?)?

   In the example, it is a problem because it may make victims feel re-traumatized. It also may make them less willing to report a crime and less willing to help with their own cases.

> **Note!**   When you write your issue statement, you should also try to include some statistics (national, state, or local) that can help further illustrate the scope of the problem.

The final part of your problem statement is your solution statement. This section will be a few sentences that describe how you think the problem can be resolved. As with the previous sections, you should think about various solutions to your problem before you draft the solution statement. For example, you could discuss a new policy or practice, or suggest a new intervention that could be created to resolve the problem.

Example:

You have concluded that a way to resolve the problems sexual assault victims encounter when they report an assault is to provide them with a professional who will help them (a victim's advocate). The advocate would be assigned to the victim when he or she first reports the crime and would help the victim through each step until the case is resolved. Your solution statement would briefly describe how the victim's advocate would help the victim feel less traumatized and would thereby improve his or her satisfaction with the system.

Once you have completed each of these steps, you will use that information to write a concise paragraph. As with all formal writing, you should write several drafts of the problem statement. You should not write only one draft.

> See **Introduction, Handout #1: "Sample Problem Statement"** at the end of this unit.

NOTES

UNIT 1
UNIT 2
UNIT 3
UNIT 4
UNIT 5
UNIT 6
UNIT 7
UNIT 8
UNIT 9

Introduction, Handout #2

Take-Home Assignment #1

(Due _____)

For questions 1, 2, and 3, print out the abstract (in landscape if necessary) and write on it the search strategy you used to find the source. In writing your search strategy, do not just list the keywords. For questions 1, 2, and 3, you must use the connectors "and" and "or" as well as parentheses and asterisks. For the first three questions, you must also use at least two synonyms for each keyword. You may use the original keywords as one of your synonyms. An example of a search strategy for "abuse of the elderly" is (abus* or harm) and (elder* or old).

1. Locate an **<u>article</u>** on **guns in schools** in *Criminal Justice Abstracts*. Print out the abstract and write on it your search strategy.

2. Locate an **<u>article</u>** on **police brutality** in *ProQuest Criminal Justice*. Print out the abstract and write on it your search strategy.

3. Locate a **<u>report</u>** on a **program that addresses violent adolescents** in **NCJRS**. Print out the abstract and write on it your search strategy.

4. Locate a **<u>report</u>** on **prostitution** in the publication database of the **National Institute of Justice**. Print out the first page of the report *and* a page from the report that shows that it is about prostitution.

UNIT
1

UNIT
2

UNIT
3

UNIT
4

UNIT
5

UNIT
6

UNIT
7

UNIT
8

UNIT
9

Introduction, Handout #2

Take-Home Assignment #2

(Due _____)

1. Draft the thesis statement for your term paper. *Failure to provide a thesis statement will result in a 5-point deduction.*

2. Locate *four* academic sources that you will use for your term paper.

   - *Two* sources must describe the program, address its history, or state its purpose(s) or goal(s);
   - *One* source must address the scope of the problem your program or policy is trying to fix; and,
   - *One* source must be an empirical study that supports your thesis statement.

   For each source, print out an abstract. Write on the abstract the topic that the source addresses. *Failure to state the topic that the source addresses will result in a deduction of one point for that source. Furthermore, if an abstract is not "on point," no points will be awarded for that source.*

UNIT 1

UNIT 2

UNIT 3

UNIT 4

UNIT 5

UNIT 6

UNIT 7

UNIT 8

UNIT 9

Introduction, Handout #2

Take-Home Assignment #3

(Due _____)

In a color (*not red*), edit the paper handed out in class on *both* a global and a local level. Make comments in the margins about problems you see with it, including organization, grammar, mechanics, and missing citations. *The more detailed your comments, the better your grade will be!* In addition, fill out the grading rubric located on the next page (add up the points for a final grade) and staple it to the front of your paper.

UNIT
2

UNIT
3

UNIT
4

UNIT
5

UNIT
6

UNIT
7

UNIT
8

UNIT
9

Student Grading Rubric for Take-Home Assignment #3				
CRITERIA	EXCELLENT 4	GOOD 3	ACCEPTABLE 2	NEEDS IMPROVEMENT 1
Global Editing:				
Ideas and Content				
Organization				
Local Editing:				
Organization				
Sentence Fluency				
Proofreading				
Citation				

Grading Scale

24 (A+)	21 (B+)	18 (C+)	15 or below (F)
23 (A)	20 (B)	17 (C)	
22 (A−)	19 (B−)	16 (C−)	

Global Editing

1. *Ideas and Content*: Is there a sufficient amount of discussion about the topic? Is it clear that the writer read and understood the research?

2. *Organization*: Did the writer organize his or her paper well on a global level? Did the writer include a thesis statement in the first paragraph? Was the content organized into logical sections? Was the paper formatted correctly (e.g., met minimum page requirement of four pages; numbered pages; used correct margin sizes)?

Local Editing

1. *Organization*: Did the writer use topic sentences? Did the writer include only one idea per paragraph? Did the writer comply with the "clumping rule"? Did the writer use transitional words to link sentences and transitional sentences to link connecting paragraphs?

2. *Sentence Fluency*: Did the writer follow the mechanics and grammar rules discussed in class? Did the writer write clearly and concisely? Are there awkward sentences?

3. *Proofreading*: Did the writer make spelling, spacing, or grammatical errors? Are there any other errors that he or she should have caught before turning in the paper?

4. *Citations*: Did the writer use complete and correct citations? Did the writer include citations everywhere they were required?

UNIT 2

UNIT 3

UNIT 4

UNIT 5

UNIT 6

UNIT 7

UNIT 8

UNIT 9

Introduction, Handout #2

Take-Home Assignment #4

(Due _____)

1. At the top of the page, write your thesis statement.

2. Create a detailed outline of your paper, which lays out a summary of the information you will present in your final paper. The outline must include *a minimum* of *ten main section heads* and *a minimum* of *two subtopics under each section head*. The outline also must follow the format taught in class (e.g., use of Roman numerals, capital letters).

 Clue!   To create your main section heads, refer to the description of the required content for the term paper listed in Introduction, Handout #3: "Writing Assignments."

3. For each *subsection head* in your outline, write the last name(s) of the author(s) and year of publication of the article(s) that you are relying on to create those subsections.

4. Submit a *minimum* of *eight abstracts* of the sources you will be using to write your paper.

UNIT 1
UNIT 2
UNIT 3
UNIT 4
UNIT 5
UNIT 6
UNIT 7
UNIT 8
UNIT 9

Introduction, Handout #2

Take-Home Assignment #5

(Due_____)

Create a reference list for your paper. Make sure that the list includes a minimum of *eight* academic sources and that it is presented in APA style, consistent with the *Publication Manual* (6th edition). You must turn in the abstract for each source. *Failure to turn in the abstracts will result in a 2-point deduction per missing abstract.*

The following grading scale will be used:

Grade	# of APA mistakes	Grade	# of APA mistakes
A+	0–2	C+	9–10
A	3–4	C	11–12
B+	5–6	D+	13–14
B	7–8	D	15–16
		F	17+

Note! Additional points will be deducted from your final score for lack of preparation or participation in the scheduled peer-review workshop for this assignment.

UNIT 1

UNIT 2

UNIT 3

UNIT 4

UNIT 5

UNIT 6

UNIT 7

UNIT 8

UNIT 9

Introduction, Handout #3

Writing Assignments

You will have two writing assignments for this class.

Assignment 1: Annotation (10% of course grade)

For this assignment, you will be required to write a one-and-a-half-page annotation of the following article. You will receive specific instructions about this assignment in a separate handout.

Etter, G. W., Sr., & Birzer, M. L. (2007). Domestic violence abusers: A descriptive study of the characteristics of defenders in protection from abuse orders in Sedgwick County, Kansas. *Journal of Family Violence, 22*, 113–119. doi: 10.1007/s10896-006-9047-x

Assignment 2: Term Paper (50% of course grade)

For the term paper, you must write a 7- to 10-page paper on one of the topics listed below. The purpose of the paper is to provide the reader with a synopsis of the policy's or program's history, purpose, goals, target population, and activities or elements; thus, you must include the following information in your paper:

- a description or definition of the program or policy, including a brief discussion of the scope of the problem (e.g., statistics) that the policy or program is trying to address;
- a discussion of the history of the program or policy, including why, where, and when it was created;
- a discussion of the purpose or goal of the program or policy (i.e., what it is designed to accomplish, who or what it is targeting); and
- a discussion of the elements of the program, including activities or actions designed to accomplish the program's goals.

In addition, you must summarize three empirical studies that have evaluated how effective the program or policy has been in achieving its goals. For each study, you must include the following information:

- a summary of the research question examined,
- a summary of the population examined,
- a summary of how the data were obtained (i.e., the type of information collected),
- a summary of the findings as they relate to the thesis statement (e.g., evidence the author[s] found that supports it), and
- a brief discussion of any significant limitations to the findings.

UNIT 1

UNIT 2

UNIT 3

UNIT 4

UNIT 5

UNIT 6

UNIT 7

UNIT 8

UNIT 9

Approved Topics for Term Paper	
Problem-Solving Courts (e.g., mental health courts, community courts)	
Boot Camps (either adult or juvenile)	Capital Punishment
D.A.R.E.	The Big Brothers/Big Sisters Program
Neighborhood Watch Programs	Prison-Based Drug Treatment Programs
Mandatory Arrest for Domestic Violence	Sex Offender Registry
Prisoner Reentry Programs	Bullying Prevention Programs

All information contained in the paper must come from academic sources; you should *not* rely on your own knowledge about the topic. Moreover, you must use a minimum of *eight* academic sources when writing your paper. Your paper must be written in APA format and must include citations written in APA format. *Failure to include citations constitutes plagiarism.* Your paper must include a title page as well as a reference page. (Neither the title page nor the reference page will count toward the 7- to 10-page requirement.) Points will be deducted from papers that do not meet the page requirement or that do not incorporate eight academic sources.

Photocopies of complete copies of *all* sources noted on the final reference list must be turned in with your paper. Papers that do not include photocopies of all of your sources will *not* be accepted. You must also submit a hard copy of the first draft of your paper in class as well as upload an electronic copy to turnitin.com. Login instructions for turnitin.com will be provided in class.

Once you turn in a complete draft of your paper, your instructor will evaluate it on its content, organization, documentation (e.g., citations), and clarity. This draft will count for 20% of your final course grade.

After receiving your instructor's evaluation, you must make the appropriate corrections to address the issues raised by your instructor and then submit a final paper; again, you will turn in a hard copy in class and upload an electronic copy to turnitin.com. With the final paper, you must again submit complete copies of the sources, and you must hand in the first draft of your paper with your instructor's comments and the grading rubric. This final paper will count for 30% of your final course grade.

UNIT 1

UNIT 2

UNIT 3

UNIT 4

UNIT 5

UNIT 6

UNIT 7

UNIT 8

UNIT 9

Introduction, Handout #4

Job Profile Assignment

The job profile assignment requires you to complete four tasks:

1. Create a résumé.
2. Interview a supervisor of a criminal justice-related organization or agency you are interested in working for in the future, and obtain information about obtaining a position with that organization or agency.
3. Transcribe the interview notes.
4. Orally present in class the information you gathered from the interview.

Task One: Create a Résumé

You must prepare a chronological résumé. It must be printed on professional-grade paper and comply with the format and style rules presented in class.

Task Two: Conduct a Job Profile Interview

You must contact a supervisor at a criminal justice-related organization or agency that you might be interested in working for in the future. You must collect information, via telephone or in person, about the specific education, skills, and experience required to compete for a position with the organization or agency. You must obtain a business card from the individual you interviewed. You must also obtain the following information:

1. Contact information for the individual you interviewed about the job. This includes *all* of the following:
   a. name,
   b. title,
   c. agency or organization,
   d. address,
   e. phone number, and
   f. email address.
2. Education:
   a. What is the minimum educational standard required for the position (e.g., associate's degree, bachelor's degree, master's degree)?
   b. What, if any, particular areas of study must a candidate have (e.g., criminal justice, ethics, business, language)?
3. Related experience (field experience):
   a. What work experience does the typical successful candidate have?
   b. What skills are particularly valuable for a candidate to have (e.g., specific training or certification)?

UNIT 1

UNIT 2

UNIT 3

UNIT 4

UNIT 5

UNIT 6

UNIT 7

UNIT 8

UNIT 9

4. Job duties:

   a. What is a general job description of the position (e.g., the type of work and activities typically required)?

   b. What are some examples of career paths within the organization? How do individuals grow within the organization over the length of their career? Is there potential for advancement?

   c. What is the average starting salary?

5. Application process:

   a. Is a background check required? If so, of what depth and type?

   b. Is any type of testing required (e.g., physical, psychological, aptitude)?

   c. Does the application process entail one interview or multiple rounds of interviews?

   d. If the applicant is required to participate in multiple rounds of interviews, does the format change at each subsequent level? If so, what are the formats?

   e. What is the average length of time between application and hire?

6. Internships:

   a. Does the organization provide internship opportunities for college students?

   b. If there are internship opportunities, how can a student secure an internship?

Task Three: Transcribe the Interview

You must transcribe your interview notes. The notes must be double spaced and written in accordance with the mechanics and grammar rules discussed in Unit 5, "Mechanics of Writing." Your notes will be graded on organization, content (Did you adequately answer *all* of the questions posed in Task Two?), and grammatical structure. The notes may be presented as a bulleted list, but all answers must be written in complete sentences. Sentence fragments or incomplete sentences are not acceptable.

Task Four: Give an In-Class Presentation

You will present to the class a summary of the information you gathered in your interview. The presentation is informal (you do not have to stand in front of the class or create slides), and you may use your transcribed notes for guidance. However, you should present the information in a conversational format, rather than reading directly from your notes.

At the end of the presentation, you will turn in your transcript, along with the business card of the individual you interviewed. You will also submit your résumé at that time.

The order in which students will present their summaries will be determined randomly. Therefore, you must be prepared to present your material on the day scheduled for presentations. If you are not prepared, this will result in a score of zero; you will not be able to make up the presentation during the next class period. Similarly, if presentations take more than one class period, you must attend both sessions. Failure to be in class on the days of presentations will result in a score of zero on the assignment.

UNIT 1
UNIT 2
UNIT 3
UNIT 4
UNIT 5
UNIT 6
UNIT 7
UNIT 8
UNIT 9

Criminal Justice Library Research

Learning Objectives

At the end of this unit, students will be able to do the following:

* Create and use a research map to find on-point articles.
* Successfully navigate several criminal justice databases and websites to find scholarly sources.

Performing Library Research: An Overview

Libraries provide many types of resources for your research. The information is available in hard copies, such as books, journals, magazines, and reports, as well as in a variety of online sources. As with any new skill, learning how to do library research takes time, practice, and patience! If you have been using search engines such as Google or Yahoo! to conduct your research, you will find that using a discipline-specific database (such as *Criminal Justice Abstracts*) will seem very difficult at first. However, after you have conducted several searches in the databases and become more familiar with them, you will find that they are fairly straightforward and easy to navigate. The bottom line is that the best way to get familiar with the various databases is to play in them!

Getting Started: Research Maps

Before you start your research, spend some time thinking about the best way to approach it. A computer will search only for the exact words you select, in the exact order you type them. Therefore, you must choose the words and their order with care so that you do not waste time reading irrelevant abstracts or articles.

Creating a Research Map

The best way to begin your library research is to create a research map. The map is based on your thesis statement and contains the keywords you will use in your computer search, in the order in which you will use them. To create a research map, follow these steps:

1. Write your thesis statement and underline the keywords.

> **Example:**
>
> List the keywords in the following thesis statement: *"Youths who use drugs are more likely to commit violent offenses."*
>
> _____
>
> _____

2. For each of your keywords, make a list of synonyms.

Example:

Write two synonyms for each of the keywords you listed in the previous example.

_____

_____

Note! If your keyword is *drugs*, you can, and should, search for certain types of drugs. This will retrieve articles that reference a particular type of drug, but that do not use the term *drug* itself.

3. Truncate all of the words that you can. To strengthen or broaden your search, you can execute a computer search for your word along with any variations of your word that an author may have used. To do this, shorten your word to the common base it shares with the words you want to find in your computer search, and add an asterisk to the end.

Note! Do not shorten your word too much. If you do, it might result in the computer using words unrelated to your search and retrieving irrelevant articles. Choose carefully the best location to place the asterisk so that the computer retrieves only relevant articles.

Example:

Truncate each of the synonyms you listed in the previous example.

_____

_____

4. Arrange your truncated synonyms into columns. To do this, write your keyword at the top of each column. Beneath each keyword, list its synonyms. This will help you visualize what you will ask the computer to find. Specifically, you will search for sources that have one of the synonyms from each column.

Example:

Create a column heading for each of the keywords you identified in the practice thesis statement. Beneath each keyword, write the synonyms you listed for that keyword. Do not forget to truncate the synonyms.

_____

_____

_____

_____

UNIT 1
UNIT 2
UNIT 3
UNIT 4
UNIT 5
UNIT 6
UNIT 7
UNIT 8
UNIT 9

5. Put your keywords together into a research map by stringing the synonyms in each column together with "or" and enclosing them in parentheses. Then connect each set of parenthetical information with "and."

Example:

Create a research map using the words you listed in the previous example.

_____

_____

Note!   If you find that you are not getting many hits, or that you are getting too many, there are several things you can do:

- Change your synonyms.
- Delete some of your synonyms.
- Check your spelling and the spacing between your words.
- Check where you have placed your asterisks.
- Use a different database.

Conducting Criminal Justice Library Research

When you conduct your library research, you must use only academic sources or scholarly journals. Many of the databases have limiters, which allow you to limit your search to those types of sources. However, when you find a source, you should always make sure it is truly academic. You can determine this by using the checklist discussed in Unit 1 (e.g., must have named authors, be longer than a few pages, include references).

There are several criminal justice databases and websites that you should use when you do your term paper research. Remember, however, that you can also find criminal justice-related sources in other disciplines such as psychology, sociology, and education. Therefore, if you are writing a paper on school violence, you should look in the education and psychology databases, as well as in the criminal justice databases. If you do use databases from other disciplines, remember to use your checklist to determine whether those articles are scholarly, too.

Criminal Justice Abstracts

This database references criminology journals and journals of related disciplines. The full texts of many (but not all) articles are available through this database. If there is not a link to the full text of an article, you can access it by other means.

When you conduct library research in the _Criminal Justice Abstracts_ database, you should begin your search in the **Advanced Search** page. There are several important features to note about this page.

- Boxes are already set up for you to enter your keywords. If you need more rows for your keywords, you can add them by clicking on the **Add Row** link.
- To the right of the keyword boxes are boxes labeled **Select a Field (optional)**. These contain a drop-down menu; click on the drop-down menu and choose the **AB Abstract or Author-Supplied Abstract** option. Other options you could choose include the last names of the

authors or the title of the article. Thus, if you have the title of the article, you can put that into the first box, click on the drop-down menu and choose the **TI Title** option to find your source.

- Beneath the search boxes is a section called **Search Options** with a list of options for you to choose from to conduct your search. The first option in this section is **Search modes**. Make sure the circle is set to **Boolean/Phrase**.
- Beneath the **Search Options** section is a section titled **Limit your results**. In that section, there is a **Linked Full Text** box. Clicking on this box returns results for full-text articles only. Do not click on this for your initial search because it might limit the number of results you get.
- After this there is a box labeled **Scholarly (Peer Reviewed) Journals**. You must use scholarly journal articles in this class, so you should check this box.
- The next box is **Publication Type**. For this class, select **Academic Journal**.
- Beneath that is a box labeled **Document Type**. For this class, choose **Article**.
- To the right of this are the **Publication Date** boxes that allow you to limit your search to the past 10 years. Using the drop-down menu, select the current month and, in the box to the right of the month, type in the current year, less 10. In the boxes beneath your starting date, use the drop-down menu to select the current month and, in the box to the right of that, type in the current year. Finally, below the **Publication Date** boxes, there is a box labeled **Language** with a drop-down menu that allows you to choose a language for your articles. Leave this box on the **All** setting.

See **Criminal Justice Library Research, Handout #1: "Library Research Reference Guide"** at the end of this unit.

Conducting Research Using a Research Map

To practice finding articles in the *Criminal Justice Abstracts* database, you will find sources for the phrase "teenagers who use drugs." Note that if you type that phrase into the first box, you will get zero hits because the computer will search for that exact phrase. Therefore, you must use a research map to find some sources.

For this exercise, use only two synonyms per keyword. For "teenagers," type teen* and youth. For "drugs" type drug* and alcohol.

1. In the first row, type (teen* or youth) into the first box. (You must type the parentheses because this database does not supply them.)
2. Change the drop-down menu on the right to **AB Abstract or Author-Supplied Abstract**.
3. Leave **AND** in the connector box.
4. Type (drug* or alcohol) into the box on the second row.
5. Change the drop-down menu on the right of that row to **AB Abstract or Author-Supplied Abstract**.
6. In the **Search modes** section, select **Boolean/Phrase**.
7. Click on **Scholarly (Peer Reviewed) Journals**. Do not click on **Linked Full Text**.
8. For **Publication Type**, select **Academic Journal**.
9. For **Document Type**, select **Article**.
10. Fill in the publication dates so that the computer will search only for articles that have been published within the past 10 years.

UNIT 1

UNIT 2

UNIT 3

UNIT 4

UNIT 5

UNIT 6

UNIT 7

UNIT 8

UNIT 9

11. In the **Language** box, select **All**.

12. Click on **Search**. This will return a list of articles. Each result has the title of the article, a list of the subjects discussed in it, a hyperlink to the full text (if available), a hyperlink to the references cited in the article, and the option to add the result to a folder.

    A hyperlink, denoted in blue, allows you to go directly to that link when you click on it. For example, when you click on the hyperlink for the full text, it will pull up the full text of the article.

13. Review the list of titles until you find an article that you think will be on point with your search for "teenagers who use drugs." Click on the title of the article. This will result in an abstract page that lists the authors, journal title, volume number, issue number, page numbers, and publication year.

14. Read the abstract carefully to make sure it is on point with your topic. If it is, click on the link to the left to view the full text of the article.

> **Note!** Not all abstracts will have links to the full text. If there is not a link to the full text, your instructor may be able to help you to determine how to obtain the full text of the article.

Conducting Research Without Using a Research Map

For your term paper, you will conduct library research for a particular program or policy that you have chosen from the list of approved topics. You will search for articles using the name of your program or policy without using a research map. If you were to use a map, you would get many unrelated hits. For example, if your topic is domestic violence courts in the United States, and you were to create the research map "(domestic or home) and (violen* or assault) and (court)," you would get hundreds of hits for articles that discuss many different types of violence, abuse, and assault that would be unrelated to your topic.

When you conduct library research for your term paper, follow these steps:

1. In the first box on the **Advanced Search** page, enter the name of your program or policy in quotation marks (e.g., "domestic violence court*"). Do not use "and" or "or."

2. Change the drop-down menu to the right to **AB Abstract or Author-Supplied Abstract**.

3. In the **Search modes** section select **Boolean/Phrase**.

4. Click on **Scholarly (Peer Reviewed) Journals**. Do not click on **Linked Full Text**.

5. For **Publication Type**, select **Academic Journal**.

6. For **Document Type**, select **Article**.

7. Fill in the publication dates so that the computer will search only for articles that have been published within the past 10 years.

8. In the **Language** box, select **All**.

9. Click on **Search**. This will return a list of articles.

10. Review the list of titles until you find one that is about domestic violence courts. When you find one that you think is on point, click on the title of the article. This will result in an abstract page that lists the authors, journal title, volume number, issue number, page numbers, and publication year.

11. Read the abstract carefully to make sure it is about domestic violence courts in the United States. If it is, click on the link to the left to view the full text of the article, or consult your instructor about how to obtain the full text of the article.

 Note! The *Academic Search Premier* database is another good database to use. Its **Advanced Search** page is similar to that of the *Criminal Justice Abstracts*. Accordingly, if you wish to do a search in *Academic Search Premier*, follow the same steps as outlined above.

ProQuest Criminal Justice

ProQuest Criminal Justice is a database with criminal justice and criminal justice-related sources. All of the sources have links to their full text.

As with *Criminal Justice Abstracts*, when you start a search in this database, you should begin on the **Advanced Search** page. There are several important features to note about this page:

1. The keyword search boxes in *ProQuest Criminal Justice* are similar to those in *Criminal Justice Abstracts*. However, in *ProQuest Criminal Justice* the box to the far right defaults to **All fields + text**. Leave the keyword search boxes set to that option. As with the *Criminal Justice Abstracts* database, you can search for a source in other ways by changing the drop-down menu.

2. Beneath the keywords search boxes are **Search options** boxes. The first two allow you to limit your search to **Full text** and **Scholarly journals**. As with *Criminal Justice Abstracts*, you should select only the **Scholarly journals** box. Do not select the **Full text** box.

3. The next option allows you to select a date range for your search. Because you must use articles published within the past 10 years, on the drop-down menu select **After this date…** Three boxes will appear. In the first box, set the drop-down menu to the current month. Set the second box to **Any Day**, and in the third box type the current year, less 10.

4. The next option allows you to select the **Source type**. Select **Scholarly Journals**.

5. Beneath that are options for **Document type**. Select **Article**.

6. The next two options are **Document feature** and **Language**. Do not select any of the boxes in those sections.

7. The next search option is **Sort results by**. Leave that box set to **Relevance**.

8. Click on **Search**.

Warning! Even though you have clicked on the box limiting the results to scholarly articles, *ProQuest Criminal Justice* will sometimes retrieve articles that are not scholarly. Therefore, it is very important that you use the checklist to determine whether the article you have found is scholarly.

Conducting Research Using a Research Map

To practice finding articles in this database, you will conduct a search typing the same keywords you used with the *Criminal Justice Abstracts* database: teen*, youth, drug*, and alcohol. To do this, follow these steps:

1. Type (teen* or youth) in the first box (you must type the parentheses).

2. Make sure the drop-down menu is set to **All fields + text**.

3. Make sure the connector box is set to **AND**.

UNIT 1

UNIT 2

UNIT 3

UNIT 4

UNIT 5

UNIT 6

UNIT 7

UNIT 8

UNIT 9

4. Type (drug* or alcohol) in the box on the second row.

5. Make sure the drop-down menu on that row is set to **All fields + text**.

6. Select **Scholarly journals** from the **Limit to** boxes.

7. Set the **Date range** box to **After this date...** by clicking on the drop-down menu.

8. Select the month and year that will result in articles published within the past 10 years.

9. From the **Source type** boxes, select **Scholarly Journals**.

10. From the **Document type** boxes, select **Article**.

11. Do not select any boxes in the options for **Document feature** or **Language**.

12. Leave the **Sort results by** box set to **Relevance**.

13. Click on **Search**.

14. This will result in a list of hits. Review the titles until you find an article that you think will be on point with your search for "teenagers who use drugs." Click on the title of the article. This will result in an abstract followed by the full text of the article.

15. Read the abstract carefully to make sure the article is on point with your topic.

16. Use the checklist to ensure that what you have found is scholarly.

Conducting Research Without Using a Research Map

Again, for your term paper, you will have to find academic sources about the program or policy you selected from the Writing Assignment handout. As previously discussed, to do this, you will not use a research map. Instead, follow these steps:

1. Type the term "domestic violence court*" in quotation marks into the first box.

2. Leave the drop-down menu set to **All fields + text**.

3. Select **Scholarly journals**.

4. Set the **Date range** boxes so that your search results are limited to the past 10 years.

5. Select **Scholarly Journals** as the **Source type** and **Article** as the **Document type**.

6. Do not select any boxes for **Document feature** or **Language**.

7. Leave the **Sort results by** box set to **Relevance**.

8. Click on **Search**.

9. This will result in a list of hits that have the phrase "domestic violence court" in their title or text. Review the list until you find an article that you think will be on point. Click on the title of the article. This will result in an abstract followed by the full text of the article.

10. Read the abstract carefully to make sure the article is on point with your topic.

11. Use the checklist to ensure that what you have found is scholarly.

National Criminal Justice Reference Service (NCJRS)

The National Criminal Justice Reference Service (NCJRS) is a federally funded agency that offers reference and referral services about crime- and justice-related issues. In part, the NCJRS website includes abstracts for criminal justice reports, articles, books, and publications. Many of the abstracts have links to the full text of the source.

To access the NCJRS Abstracts Database, follow these steps:

1. Go to www.NCJRS.gov.
2. Click on the tab at the top of the page, **Library/Abstracts**.
3. Under **Resources** click on **NCJRS Abstracts Database Search**. This will pull up a search page for the NCJRS Abstracts Database.
4. There are several ways you can find sources. If you are using a research map or keywords, you will use the **General Search** box. If you have the title of the article, the name(s) of the author(s), or the NCJ number, you will use one of the boxes specifically designated for that information.

As with *Criminal Justice Abstracts* and *ProQuest Criminal Justice*, NCJRS allows you to choose the dates you want the program to search by using the **Date Range** boxes.

Conducting Research Using a Research Map

This section will discuss how to conduct a search for "teenagers who use drugs" using a research map. To do this, use the synonyms "teen*," "youth," "drug*," and "alcohol," and follow these steps:

1. In the **General Search** box, type in the research map as follows: (teen* or youth) and (drug* or alcohol). Note that you must type the parentheses.
2. In the **Date Range** boxes, type the dates that you want to search for articles and reports, but limit this to the past 10 years.
3. Select the **All** circle next to **Choose a search type**, located in the middle of the page.
4. Click on **Search**.
5. Read through the list of articles returned; when you find one that you think might be on point, click on **Abstract**. Read it carefully to see if it is on point and from the United States. The NCJRS includes many sources that were published in other countries, so this check is important.

Conducting Research Without Using a Research Map

For your research paper, you will not use a research map. Similarly, if you want to search for a particular phrase, such as "domestic violence courts," you will do so without using a research map. To find relevant sources, follow these steps:

1. Type the name of your program or policy into the **General Search** box.
2. Select the **Phrase** circle located next to **Choose a search type** in the middle of the page.
3. Type the dates into the **Date Range** boxes so the computer searches only for sources published within the past 10 years, and click on **Search**.

Bureau of Justice Statistics (BJS)

The Bureau of Justice Statistics (BJS) is part of the U.S. Department of Justice. It produces reports of crime-related statistics and statistical trends. To access this website, go to http://bjs.ojp .usdoj.gov/.

UNIT 1
UNIT 2
UNIT 3
UNIT 4
UNIT 5
UNIT 6
UNIT 7
UNIT 8
UNIT 9

Conducting Research in BJS

There are three primary ways to find reports on this website:

1. The first way to find reports is as follows: On the top left-hand side is a tab called **Publications & Products**. When you click on this tab, the computer will open the **Publications & Products Overview** page. On that page, there are several options for you to use to continue with your research. The first is to click on the **Search** link, which will open onto a page with search boxes similar to the ones used by *Criminal Justice Abstracts* and *ProQuest Criminal Justice* (e.g., **Title**, **Author**, **Keyword**). On the **Publications & Products Overview** page, you can also conduct a search by selecting a topic from a list (e.g., **Corrections**, **Courts**) or by selecting a product type (e.g., **Publication**, **Data Table**).

2. The second way to find reports is by typing keywords into the **Enter keywords** box at the top right-hand side of the page and clicking on **GO**. When you use this box, you should limit your search to general keywords, such as "child abuse" or "gangs" rather than using a research map. After you type your keywords, click on **GO**.

3. The third way to conduct a basic search in this website is by using the menu on the left side of the page, which lists the various topics (e.g., **Courts**, **Crime Type**). If you know which topic you need to search, click on that topic.

National Institute of Justice (NIJ)

The National Institute of Justice (NIJ) is also part of the U.S. Department of Justice. In part, it produces and sponsors reports on crime and justice. To access this website, go to http://www.nij.gov/.

Conducting Research in NIJ

There are three primary ways to conduct basic research on this website:

1. Type keywords into the box that is located in the top right-hand corner of the homepage. When you use this technique, you should limit your search to general keywords, such as "child abuse" or "gangs" rather than using a research map. After you type your keywords, click on **GO**. Alternatively, you can do a more advanced search by clicking on the **Advanced Search** link under the search box. This will open a page where you can do a more specific keyword search using options such as **All of these words** and **This exact phrase**.

2. Choose from a list of topics that appears on the left-hand side of the homepage.

3. Click on the **Publications & Multimedia** tab on the top row of the homepage. On the left side of the resulting page, there are links to various types of publications, including **Recently Published** and **Publication Collections A–Z by Topic**. Each of these links brings up a list of publications or topics from which you can choose the one you are researching. You can also use the search boxes that are in the middle of the page. Type your information (title of the report, author's name, keywords) into the boxes in the section **Search for Specifics** and click on **Search**. This will result in a page with tabs that read **Published by NIJ**, **Sponsored by NIJ**, and **All Publications**. Click on the tab that relates to the type of report you are seeking.

NOTES

UNIT
1

UNIT
2

UNIT
3

UNIT
4

UNIT
5

UNIT
6

UNIT
7

UNIT
8

UNIT
9

NOTES

UNIT 1

UNIT 2

UNIT 3

UNIT 4

UNIT 5

UNIT 6

UNIT 7

UNIT 8

UNIT 9

UNIT
1

UNIT
2

UNIT
3

UNIT
4

UNIT
5

UNIT
6

UNIT
7

UNIT
8

UNIT
9

Criminal Justice Library Research, Handout #1

Library Research Reference Guide
Criminal Justice Abstracts
Conducting Research Using a Research Map

> **Example:**
>
> `(teen* or youth)` and `(drug* or alcohol)`

1. Type `(teen* or youth)` in the first box. You must type the parentheses.
2. Make sure the drop-down menu is set to **AB Abstract or Author-Supplied Abstract** in the first row.
3. Make sure the connector box is set to **AND**.
4. Type `(drug* or alcohol)` in the box on the second row.
5. Make sure the drop-down menu is set to **AB Abstract or Author-Supplied Abstract** in the second row.
6. Select the **Boolean/Phrase** circle in the **Search modes** section.
7. Select **Scholarly (Peer Reviewed) Journals**. Do *not* select **Linked Full Text**.
8. For **Publication Type**, select **Academic Journal**.
9. For **Document Type**, select **Article**.
10. Fill in the publication dates to pull up articles published within the past 10 years.
11. Set the **Language** box to **All**.
12. Click on **Search**.
13. Read through the titles until you find one that you think will be on point.
14. Click on the title.
15. Read the abstract and determine if it is on point.

Conducting Research Without Using a Research Map

> **Example:**
>
> `"DUI court*"`

1. Type `"DUI court*"` in the first box. You must use quotation marks.
2. In the drop-down menu, select **AB Abstract** or **Author-Supplied Abstract**.
3. Follow steps 6 through 15 in the previous section.

ProQuest Criminal Justice

Conducting Research Using a Research Map

Example:

(teen* or youth) and (drug* or alcohol)

1. Type (teen* or youth) in the first box. You must type the parentheses.
2. Make sure the drop-down menu is set to **All fields + text** in the first row.
3. Make sure the connector box is set to **AND**.
4. Type (drug* or alcohol) in the box on the second row.
5. Make sure the drop-down menu is set to **All fields + text** in the second row.
6. Select **Scholarly journals**. Do *not* select **Full text**.
7. Set **Date range** box to **After this date...** by clicking on the drop-down menu.
8. Fill in the publication date to retrieve articles published within the past 10 years.
9. From the **Source type** boxes, select **Scholarly Journals**.
10. From the **Document type** boxes, select **Article**.
11. Do not select any boxes from the options for **Document feature** or **Language**.
12. Make sure the **Sort results by** box is set to **Relevance**.
13. Click on **Search**.
14. Read through the titles until you find one that you think will be on point.
15. Click on the title.
16. Read the abstract and determine if it is on point.

Conducting Research Without Using a Research Map

Example:

"DUI court*"

1. Type "DUI court*" in the first box. You must include the quotation marks.
2. Make sure the drop-down menu is set to **All fields + text**.
3. Follow steps 6 through 16 in the previous section.

UNIT 1
UNIT 2
UNIT 3
UNIT 4
UNIT 5
UNIT 6
UNIT 7
UNIT 8
UNIT 9

National Criminal Justice Reference Service (NCJRS; www.ncjrs.gov)

Conducting Research Using a Research Map

Example:

```
(teen* or youth) and (drug* or alcohol)
```

1. Click on **Library/Abstracts** at the top of the page.
2. Click on **NCJRS Abstracts Database Search**.
3. Make sure the **Choose a search type** circle is set to **All**.
4. In the **General Search** box, type (teen* or youth) and (drug* or alcohol).
5. Fill in the **Date Range** boxes.
6. Click on **Search**.
7. Read through the titles until you find one that you think will be on point.
8. Click on **Abstract** and read it to determine if it is on point and from the United States.

Conducting Research Without Using a Research Map

Example:

```
DUI court*
```

1. Change the **Choose a search type** circle to **Phrase**.
2. In the **General Search** box, type DUI court*.
3. Follow steps 5 through 8 in the previous section.

UNIT 1

UNIT 2

UNIT 3

UNIT 4

UNIT 5

UNIT 6

UNIT 7

UNIT 8

UNIT 9

National Institute of Justice
(NIJ; www.nij.gov/)

There are several ways to do research on this website.

Method One

- Type your keywords into the box at the top right. Do *not* use a research map.
- Click on **GO**.

Method Two

- From the list of topics on the left side of the page, click on the one that is relevant to your search.

Method Three

- Click on the **Publications & Multimedia** tab on the top row of the homepage.
- Click on **Publication Collections A–Z by Topic**.
- Click on the topic that is relevant to your search.

Method Four

- Click on the **Publications & Multimedia** tab on the top row of the homepage.
- Type your keywords into the **Full text** box under the **Search for Specifics** section of the page.
- Click on **Search**.
- Click on the tab that relates to the type of report you are seeking.

UNIT 1
UNIT 2
UNIT 3
UNIT 4
UNIT 5
UNIT 6
UNIT 7
UNIT 8
UNIT 9

Plagiarism

UNIT SUMMARY

Learning Objectives

At the end of this unit, students will be able to do the following:

- Present an accurate definition of plagiarism and identify its most common forms.
- Avoid committing plagiarism by following general and specific rules.
- Paraphrase statistical findings into different formats (e.g., ratios, fractions, percentages, comparisons across two or more groups) to present those findings in an accurate but unique way.
- Take notes more efficiently when reviewing literature, and summarize and present that information in their own words.
- Assess what is relevant from other people's research to support their own research question or thesis statement, thereby improving their critical thinking skills.

Plagiarism: A Definition and an Overview of the Problem

Plagiarism is a form of cheating. A general definition for it is using someone else's words, ideas, statistics, or pictures and presenting them as your own. In other words, it is using another person's work without giving credit to that person. The way to give the author credit is to cite to the source from which you got the information.

Other forms of cheating include handing in a paper someone else has written for you; making up material, quotes, or sources; and buying a paper on the Internet.

Self-plagiarism is also cheating. You commit this when you hand in a paper, or portions of a paper, you wrote for another class and present that work as original.

Prevalence of Plagiarism

Plagiarism is a very real problem in academics and in the professional workforce. Studies have shown that between 26% and 54% of students admit to plagiarizing (McCabe, Trevino, & Butterfield, 2001). In the past few years, there have been several highly publicized incidents of professionals who lost their jobs as a result of plagiarizing other people's work.

The purpose of this unit is to teach you how not to plagiarize so you do not commit it in school or in your professional career.

Three Most Common Forms of Plagiarism

Plagiarism can be committed in several ways. The three most common are

1. "cutting and pasting" material from the Internet into your paper, or writing down word-for-word what you read;
2. failing to include required citations; and
3. failing to adequately paraphrase information into your own words.

Why Students Plagiarize

- *Ignorance*—Students do not know that what they are doing constitutes plagiarism.
- *Failure to understand the material*—Students find it easier to cut and paste the material than to try to understand what it means and rephrase it.

- *Poor time management*—Students wait too long to write the paper and resort to cheating to get it written on time.
- *Lack of confidence*—Students lack confidence in their research or writing abilities and are afraid to ask for help. They fear that if they turn in their own work they will get a bad grade.
- *Lack of morals*—Students find nothing wrong with plagiarizing.
- *Lack of fear of getting caught*—Students think instructors will not catch them.
- *Poor note organization*—Students fail to make a notation in their notes that the content is verbatim from the original source and requires paraphrasing. This results in their using that verbatim information in their final paper.

Rules for Avoiding Plagiarism

You can avoid the pitfalls of plagiarizing by following some very straightforward rules; some of these rules are quite specific, but we will discuss a general rule first.

General Rule

You must completely paraphrase the material and cite to the source from which you got the information. In general, paraphrasing means to rewrite the information in your own words. Citing means to put the last name(s) of the author(s) and the year the source was published somewhere in your own sentence. You can include the citation at the end of the sentence or incorporate it into the sentence text. You must have a citation for every paraphrased sentence.

Example:

Research has shown that more boys commit crimes than girls do (Wilson, 2004).

or

According to Wilson (2004), research has shown that more boys commit crimes than girls do.

We will discuss in detail how to incorporate citations into the text of your papers in the unit on APA *Publication Manual* (6th edition) rules for citing in text.

Note! You will have a lot of citations in your paper. That is okay. It is better to have too many citations than not enough. Always follow the rule, "If in doubt, cite."

Warning! Putting a citation at the end of a paragraph (as recommended by the MLA citation style) without citing throughout that paragraph is not sufficient. Under *Publication Manual* (6th edition) guidelines, placing a citation at the end of a paragraph refers only to the information in that last sentence. In such a situation, the rest of the paragraph is plagiarized.

Specific Rules for Avoiding Plagiarism

There are several specific rules you should follow so that you do not commit plagiarism.

UNIT 1 UNIT 2 UNIT 3 UNIT 4 UNIT 5 UNIT 6 UNIT 7 UNIT 8 UNIT 9

See **Plagiarism, Class Exercise #1: "Identifying Plagiarism, Part I"** at the end of this unit.

Follow these steps to determine if a "rewrite" of an original source is plagiarized:

1. Look at the rewrite and determine if it is a direct quote.

   a. If the answer is yes, check the rewrite for quotation marks, a page number, and a citation. If any of those are missing, it is plagiarized.

   b. If the answer is no, go through the rules of plagiarism guide checklist (Plagiarism, Handout #1: "Rules of Plagiarism Reference Guide"). Specifically, first determine whether every sentence is cited. Second, look for the use of synonyms, language that is too close to the original, and whether the author rewrote the example sentence by sentence. Third, determine whether the author used too many of the same words from the original source in the rewrite. If citations are missing or if the answer to any of the remaining questions is yes, the rewritten work is plagiarized.

See **Plagiarism, Class Exercise #2: "Identifying Plagiarism, Part II"** at the end of this unit.

Paraphrasing Material

When you paraphrase, you are putting information that you have read into your own words. Following are some guidelines for how to best paraphrase certain types of material.

Paraphrasing Statistics

Statistical information can be presented in many different ways that protect you from plagiarizing and that do not require direct quotes.

1. Percentages, fractions, and decimals are all interchangeable. If you want to paraphrase a statistic that includes a percentage, you can change it into a fraction or a decimal. Similarly, if you want to paraphrase a statistic that includes a fraction, you can change it into a decimal or a percentage. Finally, if you want to paraphrase a statistic that includes a decimal, you can change it into a percentage or a fraction.

Example:

How would you rewrite the statistic, 4%?

_____

2. Paraphrase statistics by approximating the numbers.

Example:

How would you rewrite 77% as an approximation?

_____

3. Paraphrase statistics by rounding them off.

Example:

How would you rewrite 2,715 as a round number?

_____

When you round off a number to paraphrase it, make sure you round to the closest and best number. For example, rewriting the statistic 2,715 as "approximately 2,750" or "approximately 2,800" is misleading and inaccurate.

 If you are rounding off a very large number, you should put the actual number in parentheses.

Example:

If the original source states that 123 million people commit a particular offense every year, and you rewrite it as "over 120 million," you should include the actual number in parentheses at the end of the sentence: (123 million). You do this so that the reader does not mistakenly believe a much higher number (i.e., 200 million) commit the offense.

Rewriting Comparisons Across Groups

When you paraphrase statistics, you may have to paraphrase information that compares two or more groups. There are several different ways to present those statistics.

Example:

You are writing a term paper about the effectiveness of drug court programs. While conducting your library research, you found that approximately 30% of the males who attended the programs reoffended, and approximately 10% of the females who attended them reoffended.

	Males	Females
Percentage who reoffended	32	10.1

There are several ways to compare these two groups:

Alternative One, write the results as a ratio: 32 divided by 10.1, which is approximately 3:1.

Alternative Two, approximate the results: "Approximately 3 out of 10 (or 30 out of 100) males reoffended, and approximately 1 out of 10 (or 10 out of 100) females reoffended."

Alternative Three, present the results as a fraction: "Approximately one-third of the males reoffended, whereas only one-tenth of the females reoffended."

(continues)

NOTES

UNIT 1

UNIT 2

UNIT 3

UNIT 4

UNIT 5

UNIT 6

UNIT 7

UNIT 8

UNIT 9

Plagiarism, Class Exercise #1

73

UNIT
1

UNIT
2

UNIT
3

UNIT
4

UNIT
5

UNIT
6

UNIT
7

UNIT
8

UNIT
9

Plagiarism, Class Exercise #1

Identifying Plagiarism, Part I

Read the following paragraph. Then state ALL of the reasons why each example constitutes plagiarism.

Original Source

Among babysitter offenses that were reported to the police, sex offenses outnumbered physical assaults 65 percent to 34 percent. . . . Most of the sex offenses involved fondling rather than the more serious crimes of rape or sodomy (41 percent, 9 percent, and 11 percent of all babysitter offenses, respectively) (Finkelhor & Ormrod, 2001, p. 3).

1. Among babysitter offenses that were reported to the police, sex offenses outnumbered physical assaults 65 percent to 34 percent.

2. Among babysitter offenses that were reported to the police, sex offenses outnumbered physical assaults 65 percent to 34 percent (Finkelhor & Ormrod, 2001).

3. "Among babysitter offenses that were reported to the police, sex offenses outnumbered physical assaults 65 percent to 34 percent" (Finkelhor & Ormrod, 2001).

4. Of the number of babysitter offenses reported, sexual offenses were greater in number than physical attacks (Finkelhor & Ormrod, 2001).

5. When babysitters commit crimes against children, they are more likely to sexually assault them than to physically harm them. Further, those babysitters are more likely to fondle the child than to rape or sodomize him or her (Finkelhor & Ormrod, 2001).

Plagiarism, Class Exercise #2

Identifying Plagiarism, Part II

Read the following paragraphs. The first paragraph is a direct quote from a research report. The next four paragraphs are "rewrites" of the first paragraph. For each paragraph, state whether it constitutes plagiarism. Then state ALL of the reasons for your answer.

Original Source

Most experts agree that any successful violence intervention program must be collaborative. Such programs should also target youth early, before frequent exposure to violence leads them to adopt negative and dysfunctional patterns of behavior (Office for Victims of Crime, 2003, p. 1).

1. Experts agree that in order for a violence intervention program to be successful, it must involve several agencies. They also agree that those programs must focus on youth when they are young, before they are exposed to violence that may cause them to engage in bad behaviors (Office for Victims of Crime, 2003).

   _____

   _____

   _____

2. According to experts, children who witness violence are more likely to benefit from violence intervention programs if they participate in them when they are young (Office for Victims of Crime, 2003).

   _____

   _____

   _____

3. Most experts are in agreement that several agencies must work together to create a successful violence intervention program (Office for Victims of Crime [OVC], 2003). They also agree that such programs should involve youth early, before constant exposure to violence causes them to use nonpositive behaviors (OVC, 2003).

   _____

   _____

   _____

4. "Most experts agree that any successful violence intervention program must be collaborative. Such programs should also target youth early, before frequent exposure to violence leads them to adopt negative and dysfunctional patterns of behavior" (Office for Victims of Crime, 2003).

   _____

   _____

   _____

UNIT 1
UNIT 2
UNIT 3
UNIT 4
UNIT 5
UNIT 6
UNIT 7
UNIT 8
UNIT 9

Plagiarism, Class Exercise #3

Rewriting Statistics

Using the spaces in the cells of the table, rewrite the following statistical information.

Alternative Ways to Write Statistics				
ORIGINAL	ALTERNATIVE #1	ALTERNATIVE #2	ALTERNATIVE #3	ALTERNATIVE #4 (OPPOSITE)
8%	.08	8/100 (8 out of 100) or 4/50 or 2/25	Almost 10%	92% do (or do not)
3/4				
.52				
Sample size: 25 males, 75 females				

Comparisons

Using the information in the first table, rewrite the statistical information in the blank cells in the second table, comparing the percentages of males versus females who report physical *abuse. Then rewrite the information comparing the percentages reporting* sexual *abuse.*

Percentage Reporting Different Types of Childhood Abuse		
	MALES	FEMALES
Physical abuse	23.1	11.6
Sexual abuse	8.7	33.9

Reported Incidents of Childhood Abuse				
	ALTERNATIVE #1	ALTERNATIVE #2	ALTERNATIVE #3	ALTERNATIVE #4
Physical abuse: Males vs. females				
Sexual abuse: Males vs. females				

UNIT 1

UNIT 2

UNIT 3

UNIT 4

UNIT 5

UNIT 6

UNIT 7

UNIT 8

UNIT 9

Plagiarism, Class Exercise #4

Paraphrasing a Paragraph From a Research Report

For this exercise, assume that your research question is whether there is a relationship between people who were abused as children and the likelihood that they will be arrested as adults for drug offenses. Then read the following paragraphs. Using the techniques you learned in class, paraphrase the information into your own words. Remember to include the relevant statistical information in your answer.

The following paragraphs are from Widom, C. S. (1995, March). *Victims of childhood sexual abuse—Later criminal consequences* (Research in Brief), p. 4. Washington, DC: National Institute of Justice.

In general, people who experience *any* type of maltreatment during childhood—whether sexual abuse, physical abuse, or neglect—are more likely than people who were not maltreated to be arrested later in life. This is true for juvenile as well as adult arrests. . . . [T]wenty-six percent of the people who were abused/neglected were later arrested as juveniles, compared with only 16.8 percent of the people who were not. The figures for adults also indicate a greater likelihood of arrest among people who were maltreated during childhood.

For certain specific offenses, the likelihood of arrest is also greater among people who were abused and/or neglected. . . . For example, 14.3 percent of the people who were abused or neglected as children were later charged with property crimes as juveniles, while this was true for only 8.5 percent of the controls. A similar difference in the rate of property crime arrests was found among adults. Childhood abuse and neglect were also associated with later arrest for drug-related offenses. More than 8 percent of the individuals abused or neglected as children were arrested for these offenses as adults, compared to only 5.2 percent of the control group.

UNIT 1

UNIT 2

UNIT 3

UNIT 4

UNIT 5

UNIT 6

UNIT 7

UNIT 8

UNIT 9

Plagiarism, Handout #1

Rules of Plagiarism Reference Guide

Rule #1: You must cite to everything that is not common knowledge.

If it is not common knowledge, you must cite to the information.

Rule #2: Do not just write your own thoughts.

All of the information in your paper must come from the sources you read. You must cite to those sources.

Rule #3: You must properly paraphrase the material.

Paraphrase means completely changing what you have read.

Rule #4: Follow the list of what NOT to do.

- Do not paraphrase one sentence at a time.
- Do not just change a few words.
- Do not just change the sentence structure.
- Do not just use synonyms.
- Do not just change the order of the sentences or just leave out words.

Doing any of the above and putting a citation at the end is plagiarism.

Rule #5: Include all of the required elements for direct quotes.

- You must use quotation marks (unless it's a block quote, 40+ words) *and*
- You must include a citation at the end of the quote *and*
- You must include a page number.

Use a direct quote only if it is said so spectacularly that you cannot paraphrase it.

BOTTOM LINE: IF IN DOUBT, CITE.

UNIT 1

UNIT 2

UNIT 3

UNIT 4

UNIT 5

UNIT 6

UNIT 7

UNIT 8

UNIT 9

Organizing a Paper: From Taking Notes to Creating an Outline

UNIT SUMMARY

Learning Objectives

At the end of this unit, students will be able to do the following:

- Effectively take notes from journal articles and research reports.
- Use those notes to create a final outline for their term paper.

Organizing an Academic Paper: An Overview

One of the most difficult parts of writing a paper is organizing the material. This is especially true when you have a large amount of information to read. However, if you organize the material as you read through each article, you will get a head start on organizing your paper.

You will write a persuasive paper for this class. In the first part of your paper, you must educate your reader about your topic. This is called the "informative section." In the second part of your paper, you will discuss three studies that have evaluated your program or policy. This is called the "persuasive section."

Taking Notes

There are certain steps you should follow to organize your information as you read through it. The first of these is to take notes for the informative part of your paper.

Taking Notes for the Informative Section of the Paper

Follow these steps when you take notes for the informative section of your paper.

1. List the topics you must discuss on a piece of paper. The topics are listed in "Introduction, Handout #3: Writing Assignments" in Unit 1; they include a description of the policy or program, the scope of the problem it is trying to address, its history, its purpose or goal, the population targeted, and its elements or activities.

2. Read the first article. As you read it and find the relevant information, highlight it but do not stop to take any notes.

3. After you have read the article, open a blank Word document. At the top, type in the last name(s) of the author(s), the year of publication, and the article's title. Also, give your source a number and type it at the top of the page. You will use that number to identify the article when you take notes.

4. Reread the article carefully, but this time, take notes about the information you must include in your paper as you read. Before each entry, type the topic you are addressing (e.g., history). After each entry, include the number you have assigned to that source and the page number where you found the information. By writing this information, you will be able to refer back to the page where you found it in the article if you need to. Also, you will use the information about each entry later when you copy and paste it into your final notes page.

UNIT 1

UNIT 2

UNIT 3

UNIT 4

UNIT 5

UNIT 6

UNIT 7

UNIT 8

UNIT 9

Example:

The first source you will take notes on is by Wilson and Carlie. You have labeled this source number "1."

Wilson & Carlie (2009) "Mental Health Courts"—source #1

History—First program created in FL in 1985 (1, p. 122)

History—Currently 250 similar programs nationwide (1, p. 124)

Purpose—To offer mentally ill offenders treatment rather than punishment and to reduce recidivism (1, p. 133)

5. After you have taken notes on your first article, follow the same rules for each remaining article, beginning a new Word document for each article.

6. After you have finished reading all of your sources and taken notes on all of the information, open a final Word document. Go back through each of the previous documents and copy and paste all of the information for the first topic into the final document.

Example:

On your notes page for the article written by Wilson and Carlie (source #1), copy all of the History information, including the parenthetical information (source number and page number) and paste it to your final document. Then go to the notes you have written for source #2, copy all of the History information from it, and move it to your final document under the Wilson and Carlie History entries. Continue to do this for all of the History sections of your notes. When you are finished with this step, you will have all of the History information from each article placed together in your final document.

7. Go back through your notes pages and copy and paste all of the information you have taken from the articles for your second topic (e.g., purpose or goal). Follow the same steps for all of the topics you have included.

8. When you have finished copying and pasting the information from each of the articles, most of the information you will need to write your paper will be in your final document. Moreover, all of the information for each element you must discuss will be grouped together in your final document. You will copy and paste that information into the appropriate sections of your outline.

Warning! When you take notes, you must completely paraphrase the information. Failing to completely paraphrase the information is plagiarism! If you do not have time to paraphrase as you take notes, put quotation marks around the information you have copied and write a large note to yourself that it is a direct copy. This will help remind you to paraphrase the material before you use it in your paper.

Taking Notes for the Persuasive Section of the Paper

After you have completed your final notes pages for the informative part of your paper, you must take notes on the three studies that you will discuss. Taking notes for this section of your paper is very similar; however, you will not combine the information from the articles as you did previously. Instead, you will take notes for each study and then use those notes to write separate discussions about each study.

To take notes for this part of your paper, follow these steps:

1. On a piece of paper, list the topics you must discuss. The topics are listed in Introduction, Handout #3: "Writing Assignments" in Unit 1; they include the research question, the population examined, how the data were obtained, the findings, and the limitations.

2. Read the first article. As you read it and find the relevant information, highlight it but do not stop to take any notes.

3. After you have read the article, open a blank Word document. At the top, type in the last name(s) of the author(s), the year of publication, and the article's title. You do not have to give it a number.

4. Reread the article carefully. This time through, write notes about the information you must include in your paper. As you read through the article, write down the information relating to each topic as you find it. Before each entry, write the topic you are addressing (e.g., data obtained, findings). After each entry, write the page number where you found the information.

> **Example:**
>
> Wilson & Carlie (2009) "Mental Health Courts"—source #1
>
> Research Question—Gender differences in recidivism rates among mental health court participants (1, p. 121)
>
> Data obtained—Sample of 60 male and 40 female program participants (1, p. 122)
>
> Data obtained—Interview and arrest records of 60 male and 40 female program participants (1, p. 122)
>
> Findings—Men more likely to recidivate than women (1, p. 130)
>
> Findings—Men ages 18–21 most likely to recidivate within the first year (1, p. 131)
>
> Findings—Women ages 18–21 least likely to recidivate within the first year (1, p. 131)

When you take notes on the findings of a study, remember that you are only interested in that part of the study that addresses your thesis statement. You do not have to take notes on irrelevant information.

 Note!   Often, researchers will discuss one element (e.g., findings) before discussing the next element (e.g., conclusions). However, some researchers intersperse the elements (e.g., discuss findings with conclusions or limitations). Be careful to write the correct topic before each entry.

After you have taken notes on all of the topics you must address in the first study, open a blank Word document. Type the title of the first study article at the top, together with the last names of the author(s) and the year of publication. Then copy and paste the information so that all of

UNIT
1

UNIT
2

UNIT
3

UNIT
4

UNIT
5

UNIT
6

UNIT
7

UNIT
8

UNIT
9

the information about each topic is grouped together. For example, all of your information about Topic 1 will be together, and your information about Topic 2 will be together. You will copy and paste that information into the appropriate section of your outline.

After you have copied and pasted the information for your first study, go back and repeat the steps for your second and third studies. When you are finished, you should have a separate document for each study.

Creating an Outline

The next step in organizing your paper is to use your notes to write an outline. The purpose of writing an outline is to further evaluate and better organize the information you have written in your notes.

In general, an outline serves as a roadmap for writing your paper. The more specific and detailed your outline, the easier it will be to write your paper. Moreover, the more detailed your outline, the easier it will be for you to stay on track and not get deterred by side issues.

General Format of an Outline

An outline has major points (major headings) and supporting points (minor headings). In an outline, major headings are denoted by Roman numerals. Minor headings are denoted by capital letters and are used to support the major headings.

Supporting points for minor headings are denoted as numbers. Supporting points for those headings are denoted by lowercase letters. If you use supporting points, you must always have a minimum of two (e.g., an "A" and a "B," a "1" and a "2," an "a" and a "b"). Do not write one without the other.

Example:

Write a sample outline using "I," "A," "B," "1," "2," "a," and "b."

_____

_____

_____

_____

_____

_____

_____

Creating an Outline From the Notes

General Rules

When you write your outline, follow these rules:

- Include the source number and page number where you found the information for each supporting point. That will enable you to easily refer back to the original source if you need to. For your studies, you need to include only the page number.
- Write the outline entries as short phrases; do not write complete sentences. In general, each entry should not be longer than one line.

Exception! The entries for your studies may be longer because they will contain more specific information. However, they should still be only one or two lines long.

- You must include an introduction and conclusion in your outline. Thus, Roman numeral I will be your introduction, and the final Roman numeral will be your conclusion.

The First Draft

When you write an outline, begin with your major points (Roman numerals). These constitute the main ideas for your paper. The major points are the elements required by your assignment. You should list them in the same order that they are listed in your assignment. For example, Roman numeral II will be "Topic 1," Roman numeral III will be "Topic 2," and Roman numeral IV will be "Topic 3."

After you have listed your major points, list your minor points. To do this, go back to your notes and copy and paste all of the information (including the source and page numbers) for the first topic you must discuss. Because you should only use short phrases in an outline, you may need to rewrite the information into short phrases.

After you have copied and pasted the information about that topic, you will need to organize it in your outline.

Example:

For your history section, you have copied and pasted information stating, "First program began in 1985 (source #, p #), started in FL (source #, p #), currently more than 250 programs in the U.S. (source #, p #)." Each piece of information will become a minor point in your outline as follows:

III. History of mental health courts

    A. Began in 1985 (source #, p #)

    B. Started in FL (source #, p #)

    C. Currently 250 similar programs nationwide (source #, p #)

After you have copied, pasted, and organized the minor points for the first major point, repeat the process for each of the other major points.

The final three major points that you must include in your outline are the three studies that have evaluated your program or policy. To create the minor points for those sections of your outline, follow the same steps that you used to create the minor points for the previous sections. Specifically, for the first study, go back to your notes and copy and paste all of the information (including page numbers) for the first point (Topic 1). Again, because you should only use short phrases in an outline, you may need to rewrite the information. After you have copied and pasted the information for the first point, repeat the process for each of the other points.

Example:

One of the topics you must discuss is the researchers' findings. For your first study, you have copied and pasted information stating, "men more likely to recidivate than women (p. #), men ages 18 to 21 most likely to recidivate within the first year (p. #), women ages 18 to 21 least likely to recidivate within the first year (p. #)." Because you have more than one finding, you must create a minor point called, "Findings" and then list each of those findings as supporting points.

VII. Study #1—Wilson & Carlie (2009)

    A. Research question—Gender differences in recidivism rates among mental health court participants

    B. Data—Interviews and arrest records of 60 male and 40 female program participants

    C. Findings

        1. Men more likely to recidivate than women (p #)

        2. Men ages 18–21 most likely to recidivate within the first year (p #)

        3. Women ages 18–21 least likely to recidivate within the first year (p #)

The Subsequent Drafts

After you have written your first draft, you must revise it. When you do the revision, concentrate on the major points, the minor points, and the overall organization of the information.

Major points.   As you read and revise your outline, consider these questions about your major points (Roman numerals):

- Do any of the major points contain so many supporting points that it would be better to split them into two or more major points?

Example:

If you have included "Purpose and goals" as one of the major points, look at how much information you have for each topic. If you find that you have a great deal, it may be better to split those topics and write them as two major points in your outline. Doing this will improve the readability and organization of your paper.

- Do any of the major points contain information either that is not very important or that should be included instead as a supporting point for another topic?

Example:

One of the topics you must discuss in your paper for this class is the population whom the program is designed to help. If you have listed each of the characteristics of the people as major points, you should rewrite them as minor points under the major point "Population served."

UNIT 1

UNIT 2

UNIT 3

UNIT 4

UNIT 5

UNIT 6

UNIT 7

UNIT 8

UNIT 9

- Did you include as major points all of the topics the assignment requires you to discuss, including the topics required for each of the three studies? If you have not, you need to add the missing information to your outline.

Minor points. After you have looked at your major points in your outline, look at the contents of your minor points. Consider these questions about your minor points:

- Would it make your paper stronger if you included the information as a major point, rather than as a minor point?

Example:

You have included "Purpose" as a supporting point for "Goals." Because these are both important points, and both have supporting information, rewrite "Purpose" as a major point in your outline.

- Are your supporting points detailed enough or have you left out important information?

Example:

You are writing a paper about drunk drivers. One of your major points is the definition of drunk drivers. You have listed gender and age as minor points. However, the literature also discusses legal limits and ethnic groups. In your outline, you should add those two minor points. In addition, determine whether there are any supporting points for those minor points (e.g., for gender you can list the supporting points "Male" and "Female").

- For sections in which you discuss the studies, did you include as minor points all of the information the assignment requires?

Example:

For your paper, the assignment requires that you discuss specific information about three studies, including the sample population, how the data were obtained, the researcher's findings, and any limitations. If you have omitted any of that information, you must add it to your outline.

The overall organization. After you have looked at the content of the major and minor points of your outline, look at the overall organization of your information. When you create your outline, you must write it so that the material makes sense, flows well, and is in a logical order.

Hint! For this class, the best way to ensure your paper is well organized is to discuss the elements in the same order in which they are listed in the assignment.

See **Organizing a Paper, Handout #1: "Sample Outline"** at the end of this unit.

NOTES

UNIT
1

UNIT
2

UNIT
3

UNIT
4

UNIT
5

UNIT
6

UNIT
7

UNIT
8

UNIT
9

Organizing a Paper, Handout #1

Sample Outline

Thesis statement: *Drunk drivers are less likely to recidivate if they receive punishment rather than treatment.*

Note! If your paper is an informative paper, you will not have a thesis statement.

I. Introduction

    A. Statistics on prevalence of drunk driving in United States (source #, p. #)

    B. Costs associated with drunk driving (source #, p. #)

    C. Two approaches to reducing drunk driving (source #, p. #)

        1. Punishment (source #, p. #)

        2. Rehabilitation (source #, p. #)

    D. Thesis statement (source #, p. #)

II. Scope of the problem

    A. Summary of most recent estimates of number of drunk driving incidents in U.S. (source #, p. #)

    B. Summary of most recent estimates of drunk driving fatalities in U.S. (source #, p. #)

III. Define drunk drivers

    A. Legal limits (source #, p. #)

        1. DUI—Minimum blood alcohol content .08 (source #, p. #)

        2. DWI—Minimum blood alcohol content .07 (source #, p. #)

    B. Prevalence—offender demographics

        1. Gender (source #, p. #)

            a. Male (source #, p. #)

            b. Female (source #, p. #)

        2. Ethnic groups (source #, p. #)

            a. Caucasian (source #, p. #)

            b. African American (source #, p. #)

            c. Hispanic (source #, p. #)

        3. Age (source #, p. #)

            a. Legal age (21 and over) (source #, p. #)

            b. Under age (under 21) (source #, p. #)

UNIT 1

UNIT 2

UNIT 3

UNIT 4

UNIT 5

UNIT 6

UNIT 7

UNIT 8

UNIT 9

IX. Study #3—Wilson, 2004

    A. Does the combined punishment of imposing a fine and revoking a DUI offender's driver's license reduce likelihood that offender will drink and drive in the future?

    B. 4000 offenders from three Midwest states

    C. Data obtained from self-report surveys, interviews

    D. Offenders who received combination of punishments less likely to drink and drive; however, if they received only one, not less likely to drink and drive

    E. Self-report survey may have led to underreporting, inaccurate answers

X. Conclusion

    A. Summary statement of problem and impact on society (source #, p. #)

    B. Restate thesis statement

    C. Future implications (source #, p. #)

        1. Create more ways to educate people about drunk driving and the repercussions for committing it (source #, p. #)

        2. Conduct more studies examining combination of punishment and treatment to reduce recidivism (source #, p. #)

UNIT 1
UNIT 2
UNIT 3
UNIT 4
UNIT 5
UNIT 6
UNIT 7
UNIT 8
UNIT 9

Mechanics of Writing: From the First Draft to the Final Paper

Proofreading Your Paper: The Final Step
    Proofreading: General Rules
    Proofreading Your Paper: A Checklist

Learning Objectives

At the end of this unit, students will be able to do the following:

- Differentiate between academic writing and other less formal styles of writing.
- Write and edit an academic paper using proper grammar, sentence structure, topic sentences, and transitions.
- Identify common grammatical and mechanical errors.
- Properly proofread a paper.

Academic Writing: An Introduction

Academic writing is a formal writing style. There are several basic rules for writing in a formal style, which you will learn about in more detail later. However, here are a few rules to help you understand what formal writing means.

1. Do not use slang or informal terminology.

Example:

Improper (informal): Cops
Proper (formal): Police officers, law enforcement officers

Example:

Improper: Teens, kids
List five proper (formal) words that you could use instead of *teens* or *kids*.

_____

_____

2. Write objectively. Do not put yourself in the paper by using first person terminology such as I, my, we, us, or our. Also, do not include personal opinion(s) in your paper.
3. Write concisely. Say what you need to say in as few words as possible; use simple rather than complex sentences.
4. Do not attempt to impress the reader by using "grandiose" language. It is easier to read and understand writing that is straightforward and to the point.

UNIT 1
UNIT 2
UNIT 3
UNIT 4
UNIT 5
UNIT 6
UNIT 7
UNIT 8
UNIT 9

5. Write as you speak (but without the slang).
6. Do not editorialize (i.e., do not preface facts or statistics with an adverb; simply state the statistic).

Example:

Improper: Tragically, 5% of all children are assaulted at some point in their lives.

Rewrite the sentence without editorializing.

_____

---

7. Do not overstate or exaggerate a fact.

Example:

Improper: It is absolutely imperative that the study be conducted.

Rewrite the sentence so that it does not exaggerate a fact.

_____

---

8. Do not use flowery or creative language. That is acceptable in creative writing, but it is not acceptable in a formal academic paper.
9. Avoid jargon (technical vocabulary or specialized words used in the profession); instead, use words and phrases more commonly used in the field. The best way to become familiar with the most common words and phrases is to read research articles and reports.

See **Mechanics of Writing, Class Exercise #1: "Identifying Mechanical Errors"** at the end of this unit.

Writing an Academic Paper: General Rules for This Class

When you write your paper, you should follow these rules:

1. Type your paper double-spaced, using a 12-point Arial, Times New Roman, or Helvetica font. Use the margins that your word processor defaults to; do not set them to accommodate the amount of material you have (too much or too little). Do not add extra line breaks between paragraphs by pressing ENTER twice.
2. Write your paper in essay format. This means you must use full sentences and correct grammar and spelling.
3. Save your paper to a disk or flash drive frequently as you work on it. Alternatively, print out what you have written at the end of each session; that way, if you have problems with your computer, you will always have a backup.

   Each time you save your paper, rename it (e.g., "version 1," "version 2"). This allows you to easily find the latest version of your work and to refer back to an earlier version if you need to.

4. Use spell check as you write your paper, but remember that it will not catch some errors (e.g., if you mistakenly type "there" instead of "their," or if you correctly spell a word, but have inadvertently typed the wrong word [e.g., you typed "sit" instead of "hit"]).

5. Use grammar check with caution. It will often make incorrect suggestions.

Writing an Academic Paper: From the First Draft to the Final Draft

Writing a paper is a process; you must write many drafts before you hand in the final one. Typically, the first draft you write will be very rough. With each subsequent draft, your paper will begin to look more like the finished product.

Getting Started: Writing the First Draft

Often, the most difficult part of writing a paper is getting started. To overcome this hurdle, follow these general rules:

1. Prepare an outline. You will use the outline as a road map to guide you as you write. It will also ensure that your paper is well organized.

2. The first time you sit down to work on your paper, just write. Do not worry about correct spelling or punctuation, or how your words sound. Just write what you want to say.

3. Do not stop and edit each sentence. Not only will you get lost in the details, but paying close attention to the little things at this stage can often lead to a discouraging lack of progress. You will go back and fix the errors in later drafts.

Writing and Editing the Second (and Subsequent) Drafts

Once you have written the gist of what you want to include in your first draft, you will go back through your paper and begin to edit it. With your second and subsequent drafts, you will read your paper with a critical eye and pay attention to the details of what you have written. Because you will need to concentrate on many details, we will discuss each of them separately.

Writing and Editing a Paper on a Global Level

You should first concentrate on the global (overall) organization of the paper. You should have an introduction, a discussion, and a conclusion. Using your outline as a guide, and the information you have written so far, start refining those elements.

The Introduction   The introduction is a short preview of your paper. Its purpose is to let the reader know what your paper will be about. If you are writing a persuasive paper, you must include your thesis statement in your introduction.

When you write your introduction, make it interesting. The more interesting your introduction is, the more likely the reader will want to continue reading. There are several ways to make your introduction interesting:

*   Think of your introduction as being similar to a movie trailer. You want to catch your readers' attention and convince them that the paper is worth reading. One way to do this is to include a relevant statistic describing the scope of the problem you are addressing.
*   Present a recent relevant media story in your introduction. By giving readers a real-world example to support your position, you pique their interest in your topic, encouraging them to read more. If you use this technique, include some background information or a brief history about your topic so that your readers better understand why you are using that example.

The Discussion   The discussion (or body) follows the introduction and is the meat of your paper. Proper organization of the discussion is very important, so be sure to follow your outline. For this class, you will write a paper about a program or policy, and you will discuss what researchers have found about its effectiveness. When you write the discussion section of your paper, assume that your readers have little to no knowledge about the program or policy you will discuss. As such, you must first educate them about it. To do this, include a brief discussion about the nature and scope of the problem that led to the creation of the program or policy, a description of the program or policy itself, a discussion of its elements, and a discussion of its purposes or goals.

For example, assume you are writing a paper about the benefits of drug courts. Your first step is to discuss the nature and scope of the problem that led to the creation of the courts.

Example:

List some things you could discuss about what led to the creation of drug courts.

_____

_____

_____

Once you have discussed the nature and scope of the problem leading to the creation of the courts, explain to your readers what a drug court is.

Example:

List some things you could discuss to explain to your reader what a drug court is.

_____

_____

_____

Finally, discuss the goals (or purposes) of drug courts.

Example:

Write two goals or purposes of drug courts.

_____

_____

Now your readers have a basic understanding about the problem(s) leading to the creation of the program or policy and about the program or policy itself. This knowledge will help your readers better understand the rest of your paper, which will include a discussion about studies that have been conducted to examine the success of the program or policy.

The Conclusion   The conclusion is a brief summary of the important points of your paper. You should not reiterate everything you have written, nor should you include new information. If you have written a persuasive paper, you should restate your thesis statement in the conclusion.

UNIT 1

UNIT 2

UNIT 3

UNIT 4

UNIT 5

UNIT 6

UNIT 7

UNIT 8

UNIT 9

If relevant, you should also suggest a course of action for the research to follow. For example, if more research needs to be conducted on certain issues pertaining to your topic, you should state that.

Writing and Editing a Paper on a Local Level

After you have looked at your paper on a global level, you must concentrate specifically on the paragraphs and sentences. This is also known as editing on a local level. When you edit on a local level, follow these general rules:

1. Include only one idea per paragraph. For example, as you write about topic A, do not go off on a tangent about topic B and then return to topic A. Write *only* about topic A.

> **Example:**
>
> If you are writing a paragraph about the daily activities of a boot camp inmate, do not mention, in the middle of the paragraph, how many boot camps exist in the United States.

   To determine whether you have written only one idea, look at each paragraph and, in the margin, write the idea that you are discussing (i.e., if you are discussing the history of your program, write *history*). If you find you have written about more than one idea, revise that paragraph.

2. Do not violate the "clumping rule." When you write, place all similar ideas together in your paper. In other words, you must group together all of your discussions about a particular issue or element.

   To determine whether you have adhered to this rule, write the topic of each paragraph in the margin of your paper. If you find that your discussion about a particular topic appears in several different places in your paper, reorganize your paper so that similar information is discussed in only one place.

3. Begin every paragraph with a topic sentence. This is a very broad statement that tells readers what the paragraph will be about. All of the following sentences in the paragraph must support that topic sentence.

> **Example:**
>
> In a paragraph in which you discuss the specific daily activities of boot camp inmates, a good topic sentence would be similar to the following:
>
> *Boot camp inmates must participate in certain daily activities.*
>
> Your subsequent sentences in that paragraph will discuss only those activities.

   To help you determine whether you have used a topic sentence for every paragraph, and whether it is the correct one to use for that paragraph, underline the first sentence of each paragraph. Then ask yourself whether the rest of the sentences in the paragraph support that topic sentence. If they do not, you must either rewrite your topic sentence or rewrite some or all of the remaining sentences in that paragraph.

UNIT 1

UNIT 2

UNIT 3

UNIT 4

UNIT 5

UNIT 6

UNIT 7

UNIT 8

UNIT 9

See **Mechanics of Writing Class Exercise #2: "Topic Sentences"** at the end of this unit.

4. Include at least three sentences in each paragraph. Fewer than that does not constitute a paragraph.

5. Use transition sentences. These connect one paragraph to the next paragraph and help a paper flow. In addition, they lead the readers from one topic to the next.

Example:

You are writing a paper about the benefits of animals in prison programs. The first paragraph states

There are many benefits to animals in prison programs. The first benefit is that they help inmates learn new skills that they can use after they are released from prison. For example, inmates who work with dogs learn how to take care of them by feeding and grooming them. They also learn basic dog-training techniques as well as the best ways to exercise a dog.

A proper transition for the next paragraph sentence would be

A second benefit of animals in prison programs is that they help to "better" the inmate. Specifically, inmates who participate in the programs develop an increased sense of self-esteem and become more confident. Moreover, because their work with the dogs will help people with disabilities, the inmates learn compassion for others.

By starting the second paragraph with a transition sentence (which also serves as the topic sentence for the paragraph), the writing flows and readers are able to easily follow the author's train of thought.

6. Use transition words. These connect the sentences within a paragraph. They are important because they help the sentences flow together and they make the writing less choppy.

See **Mechanics of Writing, Handout #1: "Transition Words and Phrases"** at the end of this unit.

There are several things you should note about this handout.

• Under the section, "To list items of information or relating to time," if you use "second ... third," you must have used "first."

Example:

Rewrite the following sentence so that it is correct:

The inmate had breakfast, second he cleaned his cell, third he went to his job.

_____

_____

Example:

Improper: In our society, I think that many youths commit crimes.

Rewrite the sentence so that it is written objectively.

_____

8. Do not editorialize. Say what you need to say without including an introductory adverb that indicates your position.

Example:

Improper: Tragically, many children are abused every day.

Proper: Many children are abused every day.

9. Write in the active rather than passive voice. Sentences that are written in the active voice begin with the person (or thing) who is doing the action.

Example:

Improper: The prison was overtaken by inmates.

Rewrite the sentence in the active voice.

_____

10. Write your sentences so that you use parallel sentence structure (e.g., between . . . and; neither . . . nor; either . . . or).

11. Do not use contractions. In other words, *don't use don't.*

12. Do not use slashes. Write out his *or* her, and he *or* she. Do not write, his/her or he/she.

Example:

Improper: An inmate is required to keep his/her cell clean.

Rewrite the sentence without the slash.

_____

13. Do not use apostrophes when writing decades and the plurals of numbers.

Example:

Improper: Researchers found errors in the first 3 data sets.

Rewrite the sentence so that it is correct.

_____

Rules About Gender

1. Be consistent in the way you use gender terms (i.e., do not write *male* and *woman* together in a sentence; use *men* and *women* or *males* and *females*).

Example:

Improper: More men were arrested than females.

Rewrite the sentence so that it is correct.

_____

2. Use the proper gender terms based on age. In particular, if you are writing about children who are younger than 12 years old, you should refer to them as *boys* and *girls*. If you are writing about teenagers between the ages of 13 and 17, you should refer to them as *male adolescents* and *female adolescents* or as *young men* and *young women*. Finally, if you are writing about people who are 18 and older, you should refer to them as *men* and *women*.

Race and Ethnicity Rules

1. When you write about racial and ethnic groups, the terms are treated as proper nouns and should be capitalized (e.g., Black, White, Asian Americans).
2. When you refer to individuals by racial groups, do not include ethnic groups in your list. In other words, do not describe some individuals by color and others by cultural heritage.

Example:

Improper: The sample population included Whites, Blacks, and Asian Americans.

Rewrite the sentence so that it is correct.

_____

3. Do not hyphenate multiword racial designations.

Example:

Improper: African-American, Native-American.

Rewrite the terms so that they are correct.

_____

UNIT 1

UNIT 2

UNIT 3

UNIT 4

UNIT 5

UNIT 6

UNIT 7

UNIT 8

UNIT 9

Commonly Confused Words

There are several words that students commonly confuse. During the editing process, if you see any of these words in your paper, stop and ask yourself whether you have chosen the correct word.

1. *Its versus it's.* *It's* always means *it is* or *it has*. It is a contraction, not the possessive form. To determine whether you have used the correct word, when you see *it's*, substitute *it is* or *it has*. If the sentence does not make sense, the correct word to use is *its*.

Example:

Improper: The shelter for homeless women lost it's funding.

Rewrite the sentence so that it is correct.

_____

2. *Between versus among.* Use the word *between* when you are referring to two people or things. Use the word *among* when you are referring to three or more people or things.

Example:

Improper: There was an altercation among the two gang members.

Rewrite the sentence so that it is correct.

_____

Example:

Improper: The four youths divided the stolen goods between themselves.

Rewrite the sentence so that it is correct.

_____

3. *Number versus amount.* Use the word *number* when you can count individual items or actions (theoretically or in reality). Use the word *amount* when you are referring to a quantity or sum.

Example:

Improper: The amount of arrests last year was higher than the previous year.

Rewrite the sentence so that it is correct.

_____

4. *Since versus because*. *Since* refers to time. If you are having difficulty remembering which word to use in your sentence, substitute the word *because* for the word "since." If it makes sense, *because* is the proper word to use.

Example:

Improper: Since the inmate assaulted the guard, he lost all of his privileges.

Rewrite the sentence so that it is correct.

_____

---

5. *i.e. versus e.g.* Both are Latin abbreviations, but they are used in different contexts: *i.e.* means *that is*; therefore, you should use *i.e.* when you want to explain or clarify your point. In contrast, *e.g.* means *for example*; you should use it when you want to strengthen your point by giving an example. When you use *i.e.* or *e.g.* within the context of a sentence, they are parenthetical; that is, you must always follow either abbreviation with a comma, and you must never begin a sentence with these abbreviations.

Example:

Improper: The offender was convicted of using a deadly weapon (e.g. a gun).

Proper: The offender was convicted of using a deadly weapon (i.e., a gun).

Improper: Many offenders commit violent crimes (i.e. rape, assault, murder).

Proper: Many offenders commit violent crimes (e.g., rape, assault, murder).

If you are confused about which term to use, substitute *in other words* for *i.e.*; if it makes sense, then *i.e.* is the correct abbreviation to use. If it does not make sense, then *e.g.* is the correct abbreviation to use.

6. *Then versus than*. *Then* refers to time (as in, next in time); *than* is used to make a comparison.

Example:

Improper: The lawyer visited his client in jail and than he posted his bond.

Rewrite the sentence so that it is correct.

_____

Improper: The new jail is larger then the old one.

Rewrite the sentence so that it is correct.

_____

---

7. *That versus who*. Use *that* when referring to anything that is not human. Use *who* when referring to people.

UNIT 1

UNIT 2

UNIT 3

UNIT 4

UNIT 5

UNIT 6

UNIT 7

UNIT 8

UNIT 9

Example:

Improper: The man that was convicted of DUI was required to attend a treatment program.

Rewrite the sentence so that it is correct.

_____

8. *Their versus they're versus there. Their* is possessive (their guns), *they're* is the contraction of *they are*, and *there* refers to a place (over there).

9. *That versus which. That* is used with a restrictive clause; *which* is used with a nonrestrictive clause. A restrictive clause (begins with "that") identifies the noun that comes before it and is essential to the meaning of the sentence. If you leave the clause out, it will change the meaning of the sentence. A restrictive clause does not require the use of commas to set it apart from the rest of the sentence. A nonrestrictive clause (begins with "which") is one that merely adds information to the sentence; it is not essential and you can leave it out. A nonrestrictive clause requires commas to set it apart from the rest of the sentence.

Example:

Improper: Rehabilitation programs, which fail to reduce recidivism, will lose their funding.

Proper: Rehabilitation programs that fail to reduce recidivism will lose their funding.

Here, the correct word is *that* because the following clause is restrictive. It identifies which programs will lose their funding and, therefore, is essential to the meaning of the sentence. Note that commas are not used to set the phrase apart from the rest of the sentence.

Example:

Improper: The prison, that was built in 1990, has both male and female inmates.

Proper: The prison, which was built in 1990, has both male and female inmates.

In this example, the correct word is *which* because the clause that follows is nonrestrictive. It does not identify the noun that comes before it, and the clause is not essential to the meaning of the sentence. It adds detail, but deleting it will not change the meaning of the sentence. Note that commas are used to set the phrase apart from the rest of the sentence.

If you are having a difficult time deciding which word to use, ask yourself if the meaning of the sentence will change if you leave the clause out. If the answer is yes, it is a restrictive clause and the correct word to use is *that*, with no commas preceding or following it. If the answer is no, it is a nonrestrictive clause, and the correct word to use is *which*, with a comma preceding it and another comma at the end of the clause.

*See **Mechanics of Writing, Class Exercise #4: "Identifying Mechanical and Grammatical Errors"** at the end of this unit.*

Proofreading Your Paper: The Final Step   119

UNIT
1

UNIT
2

UNIT
3

UNIT
4

UNIT
5

UNIT
6

UNIT
7

UNIT
8

UNIT
9

Assembling Your Final Paper: From the Title Page to the Reference List

After you have finished writing and editing your paper, and before you proofread it, you must create a title page, add a running head with page numbers, and attach the reference list.

Creating a Title Page

You must include a title page with your paper. To create a title page, follow these rules:

1. Compose a title that is a concise statement summarizing the main idea of your paper. It should be no more than 12 words and it should be formal; do not write a creative, flowery title to impress your readers. Remember—this is an academic paper.

2. Center your title and type it in the top half of the page.

3. Use the same font style and size that you use in the body of your paper; do not bold it, enclose it in quotation marks, or underline it.

4. Capitalize the first letter of each word, except for words such as *and, of, on*, and *to*. There are many rules about capitalization; for a more detailed discussion on which words to capitalize, see the *Publication Manual* (6th edition), Section 4.15.

5. Type your name, the course name or number, and the date in the center of the lower half of the page. This information must also be presented in the same font style and size that you use in the body of your paper.

Creating a Running Head With Page Numbers

Before you proofread your paper, you must insert a running head with page numbers. The running head and the page numbers should both be placed in the upper right-hand corner of each page, starting with the title page. The running head should be composed of the first four to seven words of your title and typed in all capital letters. Different computer programs require different steps for inserting running heads with page numbers; if you have questions about how to do this, click on the help button of your computer.

> See **Mechanics of Writing, Handout #3: "Sample Title Page"** at the end of this unit.

Attaching the Reference List

The final step in assembling your paper is to attach the reference list. Typically, students create the reference list as a separate document; remember to attach it to the end of your paper and to add the proper running head and page number(s) to it.

Proofreading Your Paper: The Final Step

After you have written and edited several drafts of your paper and you feel that it is finished, you must proofread it on both a global and a local level before you hand it in. This will be the final step in preparing your paper.

When you proofread your paper, you must focus on all of the guidelines you have learned in this class (e.g., organization, sentence structure), as enumerated in the next section. To help you remember what to look for as you proofread your paper, refer to the following handout, which includes all of the elements on which your paper will be graded.

> See ***Mechanics of Writing, Handout #4: "Term Paper Checklist"*** *at the end of this unit.*

Proofreading: General Rules

When you proofread your paper, follow these general rules:

1. *Proofread a hard copy.* Your eye will catch more errors if you read a hard copy than if you read it on a computer screen.

2. *Concentrate on only a few elements at a time.* When you read through your paper, you will need to look at a lot of elements. To ensure that you do not miss errors, concentrate on only a few elements at a time. Although this will require you to read through your paper several times, you will catch more errors than if you try to look for everything at once.

3. *Do not fix the errors as you find them.* As you find errors, circle them but do not stop to fix them. Doing so will cause you to lose momentum. After you have read through your paper and circled all of the errors, you will go back through and correct them.

4. *Do not wait until the last minute to proofread your paper.* Proofreading your paper properly will take time. If you rush, you will miss errors.

Proofreading Your Paper: A Checklist

When you proofread your paper, follow these steps:

☐ 1. *Proofread the overall format of the paper.* As you read through it, consider these questions:

   ☐ a. Do you have an introduction, a discussion, and a conclusion?

   ☐ b. Is your spacing correct? Are your margins the proper width (the word processor default setting), does your paper start at the top of page 1, did you hit ENTER only once between paragraphs, and do you have enough information to meet the paper's page requirement?

   ☐ c. Are your paragraphs an acceptable length? Do you have at least three sentences per paragraph, and are they not overly long?

   ☐ d. Did you number all of your pages?

☐ 2. *Proofread for sentence-level errors.* Proofread your paper by reading it out loud; your ear will catch mistakes that your eyes may miss. Circle errors as you find them, but do not stop to fix them. You will go back and fix them after you have finished reading your paper.

As you read your paper, consider these questions:

   ☐ a. Did you include your thesis statement in your first paragraph?

   ☐ b. Are your punctuation and spelling correct?

   ☐ c. Did you include only one idea in each paragraph?

   ☐ d. Did you include topic sentences for every paragraph?

   ☐ e. Did you use transition words in your paragraphs?

☐ 3. *Proofread citations.* Make sure you have properly and completely cited all sources in proper APA format, consistent with the *Publication Manual* (6th edition). Remember that you must have a citation for every sentence that contains information that is not common knowledge. If you see sentences without citations, you will need to add them. In addition, compare the citations in your paper with those on your reference list to make sure they match.

☐ **4.** *Have a third party read your paper.* After you have finished proofreading your paper, have a friend read it to see if it makes sense. (Ask someone who has good writing skills.) If something does not make sense to that person, it will probably not make sense to the instructor either, and you should revise that section. This step requires that you park your ego at the door! Do not be offended if you do not like your friend's feedback. Remember, you will make the final decision as to whether you take your friend's advice; if you do not like his or her suggestion, do not use it.

☐ **5.** *Save your work.* Save a copy of your paper to a disk or flash drive and print out an extra copy. This ensures that you will have a copy in case anything happens to your original.

> *See* **Mechanics of Writing, Class Exercise #5: "Editing Exercise"** *at the end of this unit.*

> *See* **Mechanics of Writing, Handout #5: "Grammar Reference Guide"** *at the end of this unit.*

UNIT 1

UNIT 2

UNIT 3

UNIT 4

UNIT 5

UNIT 6

UNIT 7

UNIT 8

UNIT 9

NOTES

UNIT 1

UNIT 2

UNIT 3

UNIT 4

UNIT 5

UNIT 6

UNIT 7

UNIT 8

UNIT 9

NOTES

UNIT 1

UNIT 2

UNIT 3

UNIT 4

UNIT 5

UNIT 6

UNIT 7

UNIT 8

UNIT 9

Mechanics of Writing, Class Exercise #1

Identifying Mechanical Errors

Read the following sentences. Each contains a style error. Errors may include the use of informal or emotional language, improper tone (e.g., first person), or wordiness. Rewrite each sentence so it is correct, without changing the meaning of the sentence.

1. It is evident that youths commit high numbers of crimes.

   _____

   _____

2. In the year of 1984, more men were arrested than women.

   _____

   _____

3. We, as a society, are responsible for the welfare of our kids.

   _____

   _____

4. It is imperative that we punish rapists more severely.

   _____

   _____

5. Young male juveniles have a greater number of participation in street crimes than young female juveniles.

   _____

   _____

6. Sadly, for the years between 2000 and 2008, assault was the most common type of crime committed by juveniles.

   _____

   _____

7. According to experts, forcing a child who was sexually assaulted to testify in court is totally inappropriate.

   _____

   _____

8. In the year of 1993, there were approximately 123 million episodes in which individuals got behind the wheel of a car and operated the motor vehicle while under the influence of alcohol.

   _____

   _____

   _____

   _____

UNIT 1
UNIT 2
UNIT 3
UNIT 4
UNIT 5
UNIT 6
UNIT 7
UNIT 8
UNIT 9

Mechanics of Writing, Class Exercise #2

Topic Sentences

Read the following paragraphs. Then write a topic sentence for each of them.

1. One benefit of good-time credit is that inmates' sentences are reduced, which reduces prison overcrowding. Another benefit is that inmates who receive good-time credit for attending GED classes increase their levels of education. Finally, good-time credit encourages inmates to behave, and, as a result, the number of violent acts between inmates decreases.

   _____

   _____

2. One animal rehabilitation program in prisons involves service dogs. Another program involves wild horses. A third program involves small animals, such as rabbits. Regardless of the type of animal, research has shown that animal rehabilitation programs help inmates increase their levels of self-esteem.

   _____

   _____

UNIT 1
UNIT 2
UNIT 3
UNIT 4
UNIT 5
UNIT 6
UNIT 7
UNIT 8
UNIT 9

UNIT 1
UNIT 2
UNIT 3
UNIT 4
UNIT 5
UNIT 6
UNIT 7
UNIT 8
UNIT 9

Mechanics of Writing, Class Exercise #4

Identifying Mechanical and Grammatical Errors, Part I

Part I: Look at the underlined words in each sentence and determine whether they are mechanically and grammatically correct. If they are incorrect, rewrite them so they are mechanically and grammatically correct. If they are correct, write okay *by the underlined word(s).*

1. Since the <u>1990's</u>, there has been an increase in the <u>amount</u> of teenagers who leave school before tenth grade.

   _____

   _____

   _____

   _____

2. The inmates <u>didn't</u> like the movie <u>due to the fact that</u> it was about politics, which they found to be boring.

   _____

   _____

   _____

   _____

3. <u>There is no doubt about the fact that</u> crime rates are an important topic in <u>our</u> society.

   _____

   _____

   _____

   _____

4. According to the prison administrator, "<u>Twelve</u> inmates were put into solitary confinement for fighting<u>".</u>

   _____

   _____

   _____

   _____

5. After the data <u>was</u> gathered, the researchers discovered that <u>they're</u> study was flawed.

   _____

   _____

   _____

   _____

Mechanics of Writing, Class Exercise #4

Identifying Mechanical and Grammatical Errors, Part II

Part II: Read the following sentences and determine whether they are mechanically and grammatically correct. If they are incorrect, rewrite them so they are free of mechanical and grammatical errors. If they are correct as written, write okay *by the sentence.*

1. The man was chased by the police officer.

   _____

   _____

2. At the men's detention center, after each inmate ate breakfast, they went to work.

   _____

   _____

   _____

3. It was found by researchers that three percent of all boys become truant from school.

   _____

   _____

   _____

4. Because several inmates were sick, the GED teacher cancelled their class.

   _____

   _____

   _____

5. The teens that were involved in the altercation were arrested.

   _____

   _____

   _____

6. The cop approached the man, drew his gun, and was ordering him to put up his hands.

   _____

   _____

   _____

7. Tragically, the boy who was assaulted by the man who is married to his mother, was seriously hurt.

   _____

   _____

   _____

UNIT 1
UNIT 2
UNIT 3
UNIT 4
UNIT 5
UNIT 6
UNIT 7
UNIT 8
UNIT 9

Mechanics of Writing, Class Exercise #5   **137**

UNIT 1

UNIT 2

UNIT 3

UNIT 4

UNIT 5

UNIT 6

UNIT 7

UNIT 8

UNIT 9

Mechanics of Writing, Class Exercise #5

Editing Exercise

Edit the following excerpt from a paper. You do not need to rewrite the sentences, but you do need to correct all mechanical and grammatical errors that you find. Further, if citations are missing, note where they should be included.

In 1816, Georgia law allowed for an african american to be put to death for the rape of a white woman. If a white man committed the same crime against an African American woman, his punishment was too pay a fine (Smith, 2000). Since the 1930's ninety percent of those people sentenced to death for crimes such as rape have been African American (White, 2002).

Executions were brought to a halt in the US in 1972 when the Supreme Court held that the death penalty was being given disproportionately to defendants that were African American and that most white defendants were let off the hook. Giving researchers time to research the death penalty (Jones, 2002). Ultimately, the death penalty was reinstated many years later.

In conclusion, historically the death penalty was disproportionately applied against African Americans. Poorer people were also sentenced to die more frequently than those with money (Marks, 1999). Although society has tried to fix this problem, studies show that most people executed in our county are African American. This is not acceptable!

Mechanics of Writing, Handout #1

Transition Words and Phrases

To conclude (Because of this . . . then this)

Therefore	Consequently	It follows that
As a result	Accordingly	For this reason
Hence	Thus	

Example: Drunk driving is a serious issue. _____, harsher drunk-driving laws must be passed.

To illustrate or explain an idea

For example	In particular	Specifically
To illustrate	For instance	

Example: There are many types of "street drugs." _____, one is marijuana.

To add a new point

Moreover	Further	Also
In addition	And	Furthermore

Example: All of the inmates are men. _____, they are all under the age of 21.

To list items of information or relating to time

Subsequently	Next
First (second, third)	Then

Last (requires at least two previous items in the list)

Finally (requires at least two previous items in the list)

Example: In the morning, juvenile boot camp participants first eat breakfast. _____, they clean their living areas, _____ they go to school.

To repeat or restate a similar idea

To repeat

In other words

Example: Ten percent of the offenders were female. _____, one-tenth of the offenders were female.

To indicate a conclusion

In conclusion

In sum

In summary

Example: _____, more prisons must be built in order to alleviate the overcrowding problem.

To indicate a similarity

Similarly

Likewise

Example: The leading cause of death for children is automobile accidents. _____, many young adults are also killed in car accidents.

To indicate a difference

In contrast	However
On the contrary	But

On the one hand/On the other hand

Example: Many offenders recidivate. _____ many do not.

UNIT 1
UNIT 2
UNIT 3
UNIT 4
UNIT 5
UNIT 6
UNIT 7
UNIT 8
UNIT 9

EXPOSURE TO COMMUNITY VIOLENCE 1

Exposure to Community Violence and Its Effect on Juvenile Delinquency Patterns

Jane Doe
CRJU 304
November 22, 2011

UNIT 1

UNIT 2

UNIT 3

UNIT 4

UNIT 5

UNIT 6

UNIT 7

UNIT 8

UNIT 9

Mechanics of Writing, Handout #5

Grammar Reference Guide
Sentence Structure

1. Write simple, concise sentences.
2. Avoid redundancy.
3. Avoid stilted sentences; vary your language.
4. Do not use flowery or grandiose language or exaggerate facts.
5. Do not use contractions.
6. Do not use informal language or slang.
7. Do not write in the first person (e.g., I, we, our, us).
8. Check that each noun (singular or plural) matches its pronoun.
9. Use the active voice instead of passive voice.
10. When presenting other people's research, write the findings in the past tense.
11. Check for parallel sentence structure (between … and; neither … nor; either … or).
12. Use the same verb tenses in a sentence (parallelism; e.g., The judge banged his gavel, ordered the defendant to sit down, and instructed the jury to disregard the outburst.).
13. Periods and commas go inside the quotation marks.
14. Watch the length of your paragraph—minimum three sentences, no more than three-quarters of a page.

Common Errors to Avoid

1. Watch for its versus it's. It's = it is or it has.
2. Plural dates (e.g., 1980s) do not have an apostrophe.
3. The word data is plural.
4. Know the difference: between versus among; since versus because; number versus amount; who versus that; then versus than; there versus their versus they're.
5. Do not write out percent when it is preceded by a numeral; use %.
6. Do not use slashes (e.g., he/she, and/or).
7. Use he or she for a singular pronoun, not their.
8. Write out all numbers less than 10 (general rule). Review *Publication Manual* (6th edition), Sections 4.33–4.34, for exceptions.

UNIT 1

UNIT 2

UNIT 3

UNIT 4

UNIT 5

UNIT 6

UNIT 7

UNIT 8

UNIT 9

Writing an Annotation

UNIT SUMMARY

Learning Objectives

At the end of this unit, students will be able to do the following:

- Write an annotation.
- Correctly interpret an empirical study.

Writing an Annotation: An Overview

An annotation is a short summary that explains, describes, or evaluates a source. It can vary in length from one paragraph to several pages. The purpose of an annotation is to inform the reader of the relevance, accuracy, and quality of the source cited. There are two types of annotations:

1. A descriptive annotation.

Example:

Write a definition of a descriptive annotation.

_____

_____

2. A critical annotation.

Example:

Write a definition of a critical annotation.

_____

_____

Typically, as a criminal justice student, you will not write just a single annotation. Instead, you will create an annotated bibliography, which is a bibliography with annotations for each source. There are two types of annotated bibliographies:

1. The first type simply summarizes the research that has been published on a specific topic and allows students to gain a perspective on what researchers have previously found.

2. The second type is an evaluative annotated bibliography, which presents a critical summary of the current research. When you create this type of bibliography, you evaluate each source on its own merits and compare it to, and critique it against, the other research included in your annotated bibliography.

Instructors have students create an annotated bibliography for different reasons:

1. It can provide you with the opportunity to learn more about a topic.

2. It can help you begin to frame a thesis statement. Students often have a general idea about what they would like to write their paper on, but struggle with the specific direction they would like it to take. By reading a cross-section of the relevant research, you can explore

Two Types of Commonly Annotated Criminal Justice Sources
153

UNIT
1

UNIT
2

UNIT
3

UNIT
4

UNIT
5

UNIT
6

UNIT
7

UNIT
8

UNIT
9

the different issues that have been identified in the literature. This will help you frame your thesis statement and decide which direction your paper should take.

Note! Do not confuse an annotation with an abstract. Although the two may seem similar, they are actually very different. Specifically, an annotation typically includes a description or summary of the contents of a publication. It may also include an evaluation of that content. In contrast, an abstract provides only an objective description of the contents of a publication.

Rules for Writing an Annotation

When you write an annotation, you should follow these rules:

1. Use full sentences that are grammatically correct when writing an annotation. In other words, you must follow the mechanical rules that you previously learned (e.g., write a minimum of three sentences per paragraph, use transition words, and write concisely without the slang).

2. You must write your annotation in an essay format; however, you do not have to have an introduction or a conclusion. You need introductions and conclusions only when you write academic papers and formal essays.

3. You must completely paraphrase all of the information you include in the annotation. Failure to completely paraphrase the material is plagiarism.

4. As you paraphrase the material, be careful not to change the meaning of the information.

Note! Do not change the names of the variables by using synonyms; if you do, you will change their meaning.

5. Write your annotation based on the information contained in the article. Do not rely on the information in the abstract alone.

6. Double-space the annotation, using default margins.

Two Types of Commonly Annotated Criminal Justice Sources

As a criminal justice student, the two types of sources you will most likely annotate are empirical studies and research reports.

Empirical Studies

"Empirical studies are reports of original research" (APA *Publication Manual*, 6th edition, 2010, p. 10). In them, researchers analyze either primary or secondary data, report their findings, draw conclusions, and make suggestions for future research. Empirical studies are typically published in scholarly journals.

Empirical studies have specific sections, and a good way to determine if your source is an empirical study is to look for these sections. They are the introduction, method, results (sometimes called "analysis" or "findings"), discussion, and conclusion. Each section may include some or all of the following information:

Example:

Write what each section in an empirical study contains or discusses.

Introduction:

_____

_____

Method:

_____

_____

Results:

_____

_____

Discussion:

_____

_____

Conclusion:

_____

_____

Empirical studies typically examine and analyze several variables. When you read an empirical study, you should focus only on the variables that are relevant to your thesis statement or to the topic of your paper.

Research Reports

A research report is typically published by a government agency such as the Bureau of Justice Statistics, U.S. Department of Justice, or National Institute of Justice. Like empirical studies, research reports are grounded in the research and can include either primary or secondary data. However, research reports do not always follow the format previously described for empirical studies.

Research reports include a literature review or background discussion of the subject and a description of where the data came from and how they were gathered. However, they might not include a separate discussion of the analytical techniques used to evaluate the data. Further, they often present simply a summary of the findings rather than a detailed discussion about them.

Note!   Some research reports will present the results of an outcome or evaluation study while other research reports may focus just on how a program or policy was implemented. When you conduct your library research for your paper, the three empirical studies you must find may be presented in research reports or journal articles.

Approach to Writing an Annotation

To write the most effective annotation for the handouts at the end of this unit, follow these steps:

1. Read the entire source document; do not rely only on the abstract.
2. On a separate piece of paper, or in a blank word processing document, answer the annotation questions contained in the handouts at the end of this unit. These questions direct you to the important parts of the article.

See **Writing an Annotation, Class Exercise #1: "Annotation Questions for Alcohal-Impaired Driving Article"** and **Writing an Annotation, Class Exercise #2: "General Questions for Writing an Annotation"** at the end of this unit.

3. Using your answers to the questions, write the first draft of your annotation. Remember, you must follow the mechanical and grammatical rules previously discussed.
4. Do not include information from the source that you do not understand. If you do not understand something, your paraphrase will most likely not make sense.
5. Always include the question in your answer.

Example:

The first question asks you to identify the researchers. Your answer should state, "*In this article, the researchers are Oakley and Wilson.*"

Warning!

When you answer the questions, you must completely paraphrase the information. Do not copy the answers directly from the source because that is plagiarism. If you do not have time to paraphrase the information at this point, put quotation marks around the directly quoted information and make a large note to yourself that it is a direct copy. You must go back and rewrite the information in your own words later.

6. When you write the first draft of the annotation, focus simply on getting all of the answers down on paper (or in your Word document). Do not revise and rewrite as you go.
7. After you have written the first draft, go back through and edit it following the steps for editing that you have previously learned.
8. Your final draft must be double-spaced, with default margins, and in essay format.

See **Writing an Annotation, Handout #1: "Sample Annotation"** at the end of this unit.

If your instructor assigns you the annotation writing assignment for the Domestic Violence Abusers article, see **Writing an Annotation, Writing Assignment: "Annotation Questions for Domestic Violence Abusers Article"** at the end of this unit.

UNIT 1
UNIT 2
UNIT 3
UNIT 4
UNIT 5
UNIT 6
UNIT 7
UNIT 8
UNIT 9

Writing an Annotation, Writing Assignment

Annotation Questions for Domestic Violence Abusers Article

Read the article cited below then answer the following questions. Use those answers as a guideline to write your one-and-a-half page (maximum) annotation. Be sure to write your annotation in complete sentences with proper grammar and to paraphrase completely the information in your own words. Failing to paraphrase completely or copying any information directly from the original article constitutes plagiarism. Further, all information you use must come from the text, not from the abstract.

Etter, G. W., Sr., & Birzer, M. L. (2007). Domestic violence abusers: A descriptive study of the characteristics of defenders in protection from abuse orders in Sedgwick County, Kansas. *Journal of Family Violence, 22*, 113–119. doi: 10.1007/s10896-006-9047-x

1. What type of source is this?

   _____

2. Who were the researchers?

   _____

3. How did the researchers get their data? Briefly state where the data came from.

   _____

   _____

   _____

4. What three research questions did the researchers examine?

   _____

   _____

   _____

5. What did the researchers find? You must state at least one finding that directly relates to each research question.

   _____

   _____

   _____

6. What were the limitations of the study? Specifically, what factors may have influenced the researchers' findings? You must state at least three limitations.

   _____

   _____

   _____

7. What did the researchers suggest for future research? You must state at least three suggestions.

   _____

   _____

   _____

Writing an Annotation, Class Exercise #1

Annotation Questions for Alcohol-Impaired Driving Article

Read the article cited below and then write your answers to the following questions. In class, we will use the answers as a guideline to write a one-page annotation. When you answer the questions, all of the information must come from the text, not from the abstract. Also, you must paraphrase the information in your answers; do not directly copy it from the article.

Liu, S., Siegel, P. Z., Brewer, R. D., Mokdad, A. H., Sleet, D. A., & Serdula, M. (1997). Prevalence of alcohol-impaired driving: Results from a national self-reported survey of health behaviors. *Journal of the American Medical Association, 277*(2), 122–125.

1. What type of source is this?

   _____

2. Who were the researchers and what was the purpose of their study?

   _____

   _____

   _____

3. How did they get their data? Briefly state where the data came from.

   _____

   _____

   _____

4. What did the researchers find? In other words:

   a. What did they find regarding the number of participants who drove while impaired by alcohol?

   _____

   _____

   b. What generalizations did they make based on that number?

   _____

   _____

   c. Which participants did they find were the most likely to drink and drive?

   _____

5. What did the researchers conclude from their findings?

   _____

   _____

   _____

6. What were the limitations of the study? Specifically, what factors may have influenced the researchers' findings?

_____

_____

_____

7. What recommendations did the researchers make to help reduce the number of incidents of drunk driving?

_____

_____

_____

UNIT 1

UNIT 2

UNIT 3

UNIT 4

UNIT 5

UNIT 6

UNIT 7

UNIT 8

UNIT 9

Writing an Annotation, Class Exercise #2

General Questions for Writing an Annotation

Read the assigned article and then write your answers to the following questions. In class we will use the answers as a guideline to write an annotation. When you answer the questions, all of the information must come from the text, not *from the abstract. Also, you must paraphrase the information in your answers; do not directly copy it from the article.*

1. What type of source is this?

   _____

2. Who were the researchers?

   _____

   _____

3. What was the purpose of the study (what were the research questions)?

   _____

   _____

   _____

4. How did the researchers get their data?

   _____

   _____

5. What did the researchers find?

   _____

   _____

6. What did the researchers conclude from their findings?

   _____

   _____

7. What were the limitations of the study? Specifically, what factors may have influenced the researchers' findings?

   _____

   _____

   _____

8. What did the researchers suggest for future research?

   _____

   _____

   _____

UNIT 1

UNIT 2

UNIT 3

UNIT 4

UNIT 5

UNIT 6

UNIT 7

UNIT 8

UNIT 9

Writing an Annotation, Handout #1

Sample Annotation

Liu, S., Siegel, P. Z., Brewer, R. D., Mokdad, A. H., Sleet, D. A., & Serdula, M. (1997).
   Prevalence of alcohol-impaired driving: Results from a national self-reported survey of
   health behaviors. *Journal of the American Medical Association, 277*(2), 122–125.

   In this empirical study, Liu et al. used data obtained from a 1993 telephone survey to
estimate the number of times adults drink alcohol and then drive. In doing so, they analyzed
data collected by the Behavioral Risk Factor Surveillance System (BRFSS) in 49 states and
the District of Columbia. They found that 2.5% of the people who participated in the study
had driven while intoxicated in the previous month. From that percentage, they determined
that 123 million instances of drunk driving had occurred that year. In addition, they found
that men were almost 5 times more likely to drive while under the influence of alcohol than
women were. They also found that young men, non-Hispanic Whites, non-White women, and
binge drinkers were most likely to drink and drive. They concluded that from 1986 through
1993, there was a 20% decline in the number of drunk-driving incidents in the United States.
They also concluded that from 1982 through 1992, there was a 30% decrease in the number of
drunk-driving-related fatalities.
   The researchers noted several limitations to this study. First, they stated that people without
telephones and people under the age of 18 were not included in the survey. They also noted
that people might have underreported the number of times they drank and drove. To reduce
the number of individuals who drink and drive, researchers recommended harsher sanctions
and the involvement of clinicians to counsel people with drinking problems.

UNIT 1

UNIT 2

UNIT 3

UNIT 4

UNIT 5

UNIT 6

UNIT 7

UNIT 8

UNIT 9

Creating a Reference List in APA Style

Learning Objectives

At the end of this unit, students will be able to do the following:

- Identify the difference between a bibliography and a reference list.
- Prepare a reference list in APA style, in accordance with the *Publication Manual* (6th edition).

APA Reference List: An Overview

There are several important introductory facts about reference lists:

- The purpose of a reference list is to give readers information about the source so they can locate a copy of it.
- A reference list is not the same as a bibliography.

Example:

Write the difference between a reference list and a bibliography.

_____

_____

- Every source you cite in the text of your paper must be listed on your reference list. Similarly, every source you list on your reference list must be cited in your paper.

There are many *Publication Manual* rules about citing on the reference list. You will learn how to cite to the most commonly used sources. For all other citation types, or citing questions you may have, refer to the *Publication Manual* (6th edition). You can also go to www.apastyle.org/learn /index.aspx for further information about citing in APA style.

Citations on the Reference List

General Formatting Rules

1. Center the title at the top of the page in the same size and style of font as used in the rest of your paper.
2. The correct title is References, not Reference List, Bibliography, Reference, or Works Cited. (Works Cited is used only for an MLA-style reference list.)
3. Capitalize only the first letter of References. Do not bold, italicize, or underline the title.
4. Do not number the entries on the reference list.
5. Double-space the reference list entries. Do not add an extra line space between each source entry.
6. Begin each entry with a hanging indent. To do this, begin each source flush with the left-hand margin. When you are finished typing all of the information for the citation, format the entry so the second line of text and all subsequent lines for that entry are indented. (This is most easily done by formatting the entire list after all the entries have been typed.)

Example:

Lee, J. M., Steinberg, L., & Piquero, A. R. (2010). Ethnic identity and attitudes toward the police among African American juvenile offenders. *Journal of Criminal Justice, 38*(4), 781–789.

7. Insert a single space between each of the elements in a citation (e.g., between the last author's initial and the date).
8. Present the sources alphabetically by the first author's last name.
9. Do not change the order of the authors within a citation on the reference list. Include them on the reference list in the same order in which they are listed in the original source.
10. In general, most sources you will cite to contain the same elements. These are the author(s), year of publication, title, and publisher information or data required to retrieve the source.

Citing to the Authors

The first element of a citation on the reference list is the name(s) of the author(s). When you write a citation, always get the name(s) of the author(s) from the original source and not from the abstract in the database; the abstract does not always list all of the authors, especially when there are more than three. If you rely on the abstract for the name(s) of the author(s), your reference list may be incorrect.

Citations With One Author

Write the last name followed by the first initial of the author's first name, then a middle initial if one is given. If the author's name includes a Jr. or a number (e.g., III), you must include it. However, do not include credentials such as JD or MD.

Example:

Write a citation for an article written by K. W. Wilson.

_____

Write a citation for an article written by K. W. Wilson Jr.

_____

Citations With Two Authors

If there are two authors, list both and separate the names with an ampersand (&). Include a comma before the ampersand.

Example:

Write a citation for an article written by L. M. Wilson and S. K. Oakley.

_____

UNIT 1
UNIT 2
UNIT 3
UNIT 4
UNIT 5
UNIT 6
UNIT 7
UNIT 8
UNIT 9

Citations With Three to Seven Authors

If there are three to seven authors, you must list all of the authors and put an ampersand before the name of the last author. Include a comma before the ampersand.

Example:

Write a citation for an article written by T. Oakley, R. Jackson, L. S. Marvin, and L. M. Summers.

_____

Citations With Eight or More Authors

If the citation has eight or more authors, list the first six authors followed by three ellipsis points and then the last author's name. Include a comma before the ellipses. There is no ampersand before the last name.

Example:

Write a citation for an article written by R. Taylor, T. Thomas, L. Jackson, T. Oakley, W. Wilson, K. L. Harvey, R. Charles, and Y. Roger.

_____

Special Circumstances When Citing to Two or More Authors

Authors often publish articles with different coauthors. If you want to cite to those sources, alphabetize by the second author's last name:

Example:

You have two articles; one was published by Y. D. Wilson and E. B. Bryce, and the other one was published by Y. D. Wilson and L. F. Wells. Write the two entries as they would appear on the reference list.

_____

_____

If the second author is the same, alphabetize by the third author's last name, and so on.

Citing to the Date of Publication

The next element after the author(s) is the date of publication. For this element, write the year of publication in parentheses. Do not include a month or season, even if one is given.

Citing to a Single Author With Works Published in Different Years

Often, an author will have published several articles. If you want to cite to more than one of his or her articles, cite to the oldest work first.

UNIT
1

UNIT
2

UNIT
3

UNIT
4

UNIT
5

UNIT
6

UNIT
7

UNIT
8

UNIT
9

Example:

J. P. Wilson published an article in 2000 and another one in 2001. Write how the entries would appear on the reference list.

_____

_____

Citing to a Single Author With Several Works Published in the Same Year

Occasionally, an author may have several articles published in the same year. If this occurs and you want to use more than one article, you must look to the title of the article to determine the order of the sources on the reference list.

Example:

In 2000, Wilson published an article titled "An Analysis of Truancy Recidivism Data." That same year, he published an article titled "An Evaluation of a School Truancy Program." Because the word *analysis* begins with an "a" and the word *evaluation* begins with an "e," the article entitled, "An Analysis of Truancy Recidivism Data" will be listed first on the reference list.

When you list these sources on your reference list, you must use lowercase letters next to the publication year to differentiate them.

The entries would be listed on the reference list as follows:

Wilson, T. (2000a). An analysis of truancy recidivism data.

Wilson, T. (2000b). An evaluation of a school truancy program.

Citing to the Title of a Journal Article

In a citation to a journal article, the next element after the date of publication is the article title. The title is typed in regular font (no italics). The only capitalized words in the title are the first word, any proper nouns, and the first word of a subtitle.

Example:

Article title with no subtitle: Understanding plagiarism among college students.

Article title with a subtitle: Understanding plagiarism among college students: Reasons why students plagiarize.

Note! Often an article title will be capitalized in the original source. Even if this is the case, follow the *Publication Manual* (6th edition) rules regarding the proper way to cite to the title. Do not follow the format of the original source.

Citing to the Title of a Journal

In a citation to a journal article, the next element after the article title is the journal title. The journal name is typed in italics. Each word is capitalized except for words like *of*, *the*, *on*, and *in* (unless they are the first word of a subtitle).

Example:

Journal of Criminal Justice

Citing to the Volume and Issue Numbers

In a citation to a journal article, the next element after the journal title is the volume number of the article. The volume number is also typed in italics.

Example:

Journal of Criminal Justice, 42

Note that there is a comma after the title and a space before the volume number.

After the journal volume number, cite to the issue number if one is given. Not all sources will have an issue number, but if one is given, you must include it. The issue number is typed in regular font (no italics) and enclosed in parentheses. There is no space between the volume number and the issue number.

Example:

Journal of Criminal Justice, 42(2)

Citing to the Pages of a Journal Article

The final element of a citation to a journal article is the page numbers of the article. You must include both the first and last page numbers. Do not write the word *pages* or the abbreviation *pp*. Instead, type only the numbers. Often the abstract included in the database where you found your source will not include the last page number, so you must obtain the first and last pages from the original source. If you list only the first page, your citation will be incomplete.

Example:

Journal of Criminal Justice, 42(2), 223–242.

The Finished Citation

Using the elements discussed, the final citation is:

Example:

Wilson, K. W. (2003). Understanding plagiarism among college students. *Journal of Criminal Justice, 42*(2), 223–242.

See the following handouts at the end of this unit:

Creating a Reference List in APA Style, Class Exercise #1: "Identifying APA Reference List Errors"

Creating a Reference List in APA Style, Handout #1: "Different Types of Sources"

Creating a Reference List in APA Style, Handout #2: "Sample Reference List"

UNIT 1

UNIT 2

UNIT 3

UNIT 4

UNIT 5

UNIT 6

UNIT 7

UNIT 8

UNIT 9

NOTES

UNIT
1

UNIT
2

UNIT
3

UNIT
4

UNIT
5

UNIT
6

UNIT
7

UNIT
8

UNIT
9

NOTES

UNIT 1

UNIT 2

UNIT 3

UNIT 4

UNIT 5

UNIT 6

UNIT 7

UNIT 8

UNIT 9

Creating a Reference List in APA Style, Class Exercise #1

Identifying APA Reference List Errors

Read through the following citations and determine what types of sources they are (e.g., journal article, book). Then determine whether the underlined words and numbers are written in correct APA format. If they are correct, write correct *above the underlined phrase. If they are incorrect, write the corrected version on the lines below the entry. After you have finished correcting the citations, look at the order of the citations. Are they in the correct order in which they should appear on a reference list? If not, write the last name of the first author in each entry in the correct order on the six numbered lines at the end of the handout.*

Reference

_____

<u>Chang, D. C., & Cornwell, E. E., & Sutton, E. R.</u> (2005). <u>*A multidisciplinary youth violence prevention initiative: impact on attitudes*</u>. *Journal of the American College of Surgeons, 201*(5), <u>721</u>.

_____

_____

_____

<u>Parent, David G. (2003, June).</u> <u>Correctional Boot Camps: Lessons from a decade of research</u>. <u>Washington, D.C.</u>: National Institute of Justice.

_____

_____

_____

Farrer, B. A. (1998). Changing professional ideology in the United States. In B. K. Schwartz & <u>H. R. Jones</u> (<u>eds.</u>), *The sex offender: New insights and treatment developments* (<u>pages</u> 21–35). Kingston, NJ: Civic Research Institute.

_____

_____

_____

<u>Rose, E. & James, C.</u> (<u>2003</u>). *Restorative justice and the legal model*. Retrieved <u>on June 30, 2004</u> from http://www.vera.org/publications

_____

_____

_____

UNIT 1
UNIT 2
UNIT 3
UNIT 4
UNIT 5
UNIT 6
UNIT 7
UNIT 8
UNIT 9

Chesney-Lind, M., & Paramore, V. V. (2001). <u>Are girls getting more violent? Exploring juvenile robbery trends</u>. *Journal of Contemporary Criminal Justice,* <u>17(2)</u>, 142–166. <u>DOI: 10.1177/1043986201017002005</u>

_____

_____

_____

Smith, J. (2000). <u>Introduction to Legal Philosophy</u>. <u>Miami Beach, FL:</u> Sage.

_____

_____

_____

1. _____

2. _____

3. _____

4. _____

5. _____

6. _____

UNIT 1

UNIT 2

UNIT 3

UNIT 4

UNIT 5

UNIT 6

UNIT 7

UNIT 8

UNIT 9

Creating a Reference List in APA Style, Handout #1

Different Types of Sources

Print Sources

Journal Article

Lee, J. M., Steinberg, L., & Piquero, A. R. (2010). Ethnic identity and attitudes toward the police among African American juvenile offenders. *Journal of Criminal Justice, 38*(4), 781–789.

Research Report

McCord, J., & Conway, K. P. (2005). *Co-offending and patterns of juvenile crime* (Research in Brief). Washington, DC: National Institute of Justice.

Book

Collins, R. (2008). *Violence: A micro-sociological theory.* Princeton, NJ: Princeton University Press.

Article or Chapter in an Edited Book

Lehti, M., & Aromaa, K. (2006). Trafficking for sexual exploitation. In M. Tonry (Ed.), *Crime and justice: A review of research* (pp. 133–227). Chicago: University of Chicago Press.

Electronic Sources

Article From a Website

Shah, S., & Estrada, R. (2009). *Bridging the language divide: Promising practices for law enforcement.* Retrieved from http://www.vera.org/download?file=111 /Engaging%2Brespondents.pdf

Article From a Database With a DOI Number

Dembo, R., & Gulledge, L. M. (2009). Truancy intervention programs: Challenges and innovations to implementation. *Criminal Justice Policy Review, 20*(4), 437–456. doi: 10.1177/0887403408327923

Article From a Database Without a DOI Number

Hennessy, D. A., & Wiesenthal, D. L. (2004). Age and vengeance as predictors of mild driver aggression. *Violence and Victims, 19*(4), 469–477. Retrieved from http://www .springerpub.com/product/08866708

UNIT 1

UNIT 2

UNIT 3

UNIT 4

UNIT 5

UNIT 6

UNIT 7

UNIT 8

UNIT 9

Creating a Reference List in APA Style, Handout #2

Sample Reference List

References

Alpert, G. P., & Huff, C. R. (1983). Defending the accused: Counsel effectiveness and strategies. In W. E. McDonald (Ed.), *The defense counsel* (pp. 247–271). Beverly Hills, CA: Sage.

Bayari, C. (2003). Sentencing drink-driving offenders in the NSW local court. *Sentencing Trends & Issues, 27*, 1–18.

Beck, J. C., & Shumsky, R. (1997). A comparison of retained and appointed counsel in cases of capital murder. *Law and Human Behavior, 21*(5), 525–538.

Bertelli, A. M., & Richardson, L. E., Jr. (2007). Measuring the propensity to drink and drive. *Evaluation Review, 31*(3), 311–337. doi: 0.1177/0193841X070310030401

Breckenridge, J. F., Winfree, L. T., Maupin, J. R., & Clason, D. L. (2000). Drunk drivers, DWI "drug court" treatment, and recidivism: Who fails? *Justice Research and Policy, 2*(1), 87–105.

Grohosky, A. R., Moore, K. A., & Ochshorn, E. (2007). An alcohol policy evaluation of drinking and driving in Hillsborough County, Florida. *Criminal Justice Policy Review, 18*(4), 434–450. doi: 10.1177/0887403407303736

Homant, R. J., Kennedy, D. B., & Evans, W. C. (2007). Evaluating last call: A program directed at outstanding drunk driving warrants. *Police Quarterly, 10*(4), 394–410. doi: 10.1177/1098611107307735

UNIT 1

UNIT 2

UNIT 3

UNIT 4

UNIT 5

UNIT 6

UNIT 7

UNIT 8

UNIT 9

Citing in the Text in APA Style

Citing to Two Authors

    Citing to the Same Source With Two Authors Within a Paragraph

    Citing to the Same Source With Two Authors in Subsequent Paragraphs

Citing to Three, Four, or Five Authors

    Citing to the Same Source With Three to Five Authors Within a Paragraph

    Citing to the Same Source With Three to Five Authors in Subsequent Paragraphs

Citing to Six or More Authors

    Citing to the Same Source With Six or More Authors Within a Paragraph

    Citing to the Same Source With Six or More Authors in Subsequent Paragraphs

Combining In-Sentence Citations With Citations at the Ends of Sentences

Miscellaneous Rules for Citing in the Text

    Citing to a Secondary Source

    Incorporating a Quote Into a Sentence

Learning Objectives

At the end of this unit, students will be able to cite in the text in APA style in accordance with the APA *Publication Manual* (6th edition).

Citing in the Text in APA Style: An Overview

When you cite in the text of your paper, you must do so in accordance with the *Publication Manual of the American Psychological Association, Sixth Edition*. This unit will teach you how to do that. There are several important introductory facts and rules about citing in the text in APA style:

1. There are two purposes for citing in the text.

Example:

Write the two purposes for citing in the text.

_____

_____

---

2. You must cite to all information that is not common knowledge.

UNIT 1

UNIT 2

UNIT 3

UNIT 4

UNIT 5

UNIT 6

UNIT 7

UNIT 8

UNIT 9

Example:

Write a definition for *common knowledge*.

_____

3. Every source you cite to in your paper must be included on your reference list and vice versa.

4. Every citation must contain the same basic information: the last name(s) of the author(s) and the year of publication.

5. There are three main ways you can cite in the text:

   a. Provide a citation at the end of the sentence.

   b. Incorporate a citation into the sentence.

   c. Blend the two by citing at the end of some sentences and incorporating the citations into the text of others.

   Although all of these are acceptable ways to cite, blending your citations is the preferred method because it adds variety to your paper and makes it flow more smoothly.

There are many APA rules about citing in the text. This unit will cover the basic rules that you are most likely to use when you write your paper for this class. For additional rules about citing in the text, refer to the *Publication Manual* (6th edition) or www.apastyle.org/learn/index.aspx.

Citing to a Source at the End of a Sentence

Placing a citation at the end of a sentence is the most basic way to cite to a source. To do this, include the last name(s) of the author(s) and the year of publication in parentheses at the end of the sentence.

General Rules

When you cite to a source at the end of a sentence, follow these rules:

- Write only the last name(s) of the author(s). Do not include first names or initials, and do not include a person's qualifications (e.g., JD, MD).
- When there are two or more authors of a source, list their names in the same order that they are listed in the original source. Never change the order of the authors.
- Place the period after the citation because it is part of the sentence.
- Include only the year of publication; do not include a month or season, even if one is given in the original source.
- When you cite at the end of the sentence, it means that everything you said *in that sentence* is attributable to that author. However, if you place it at the end of a paragraph, it means that *only that last sentence* is attributable to that author. In that situation, if any of the remaining sentences in the paragraph do not have citations, you will have committed plagiarism.

Citing to One Author

When you cite to one author, include the last name and the year of publication in parentheses. Place a comma between the name and the year of publication.

Example:

Smith published an article in 2000. Write the citation that would come at the end of the following sentence: *Most crimes are committed by males.*

_____

Every time you cite to a single author at the end of a sentence, you must include both the last name and the year of publication.

Citing to Two Authors

When you cite to two authors, include the last names and the year of publication in parentheses. Separate the names by an ampersand (&) and place a comma before the year of publication.

Example:

Smith and Jones published an article in 2001. Write the citation that would come at the end of the following sentence: *Boys commit more crimes than girls do.*

_____

Every time you cite to two authors at the end of a sentence, you must include both names and the year.

Example:

Smith and Jones published an article in 2001. Rewrite the following two sentences with the correct citations: *Boys commit more crimes than girls do. Most of the crimes they commit are misdemeanors.*

_____

_____

This rule also applies when there is an intervening citation.

Example:

Boys commit more crimes than girls do (Smith & Jones, 2001). The majority of these crimes are misdemeanors (Otis, 2009). However, some boys do commit more serious crimes, such as assault and burglary (Smith & Jones, 2001).

Citing to Three, Four, or Five Authors

When you cite to three, four, or five authors, include the last names and the year of publication in parentheses. Separate the names by commas, place a comma and then an ampersand before the last name, and place a comma before the date of publication.

UNIT 1
UNIT 2
UNIT 3
UNIT 4
UNIT 5
UNIT 6
UNIT 7
UNIT 8
UNIT 9

Example:

Oakley, Wilson, and Jones published an article in 2001. Write the citation that would come at the end of the following sentence: *Boys commit more crimes than girls do.*

_____

The first time you cite to three, four, or five authors in your paper, you must include all of the names, followed by the year of publication. Every subsequent time you cite to that source in your paper, write the citation as (Oakley et al., 2001); et al. is a Latin abbreviation for *et alia*, which means *and others*. Note that there is no period after the *et* but there is one after the *al.* Note also that in the citation, there is no comma after Oakley but there is one before the year.

Example:

Oakley, Wilson, and Jones published an article in 2001. Rewrite the following sentences with the correct citations: *Boys commit more crimes than girls do. However, most of the crimes they commit are misdemeanors.*

_____

_____

This rule also applies when there is an intervening citation.

Example:

Boys commit more crimes than girls do (Oakley, Wilson, & Jones, 2001). The majority of these crimes are misdemeanors (Otis, 2009). However, some boys do commit more serious crimes, such as assault and burglary (Oakley et al., 2001).

Citing to Six or More Authors

When there are six or more authors, every time you cite to that source, write just the first author's last name, followed by et al. and a comma, and then the year of publication.

Example:

Wilson and seven other authors published an article in 2001. Write the correct citation for the end of the following sentence: *Boys commit more crimes than girls do.*

_____

Citing to a Government Agency

When the author is a government agency, the first time you cite to the source put the name of the organization followed by the accepted abbreviation in brackets, then the year of publication. Every subsequent time you cite to the source, write just the abbreviation and year of publication.

Example:

The first drug courts were established in the late 1980s (National Institute of Justice [NIJ], 2006). By the end of 2005, there were over 1,500 drug courts in the United States (NIJ, 2006).

Citing to Multiple Sources ("String" Citations)

Sometimes you find the same information stated in several articles. To cite to all of the sources, you must list the authors alphabetically by first author and separate the sources by semicolons. After each author's name, write a comma and the year of publication.

Sources With Only One Author

When you write a string citation with sources by one author, list the authors alphabetically, not chronologically.

Example:

Most crimes are committed by males (Oakley, 2003; Smith, 2001; Wilson, 2000).

Sources With Two, Three, Four, or Five Authors

When a source has more than one author, alphabetize that source in the string citation by the first author's last name. In addition, when a source has two authors, separate the names within the source by an ampersand. When a source has three to five authors, separate the authors' names within the source by commas, and place an ampersand before the last author's name.

Example:

Men are more likely to drink and drive than women are (Carson, 2003; Clover & Otis, 2011; Jackson, Barnwell, & Halley, 2005).

When a source has three to five authors and you have already cited to that source in your paper, you do not need to list all of the authors again. Instead, write the first author's name followed by et al., then a comma, and then the year of publication.

Example:

Young men between the ages of 18 and 21 are more likely to drink and drive than older men are (Carson, 2003; Clover & Otis, 2011; Jackson et al., 2005).

Sources With Six or More Authors

When a source has six or more authors, every time you cite to the source, write only the first author's last name followed by et al., then a comma, and then the year of publication.

> **Example:**
>
> Men are more likely to drink and drive than women are (Carson et al., 2003; Clover & Otis, 2011; Jackson, Barnwell, & Halley, 2005; Parker, 2006).

Sources With an Author as a Single Author and as a Coauthor

If an author has published two articles, one with another author and one without, cite first to the article in which he or she is the sole author.

> **Example:**
>
> Men are more likely to drink and drive than women are (Clover, 2009; Clover & Otis, 2008).

Sources With the Same Author, Published in Different Years

If an author has published articles in different years, cite to the oldest work first. Separate the years of publication by a comma.

> **Example:**
>
> Men are more likely to drink and drive than women are (Darnell, 1997, 2001).

Sources With the Same Author, Published in the Same Year

If an author has published more than one article in the same year, you must look at the titles of the sources to determine how to cite to them. The source with the title that comes first alphabetically will be listed with an "a" following the year of publication. The source that comes second alphabetically will be listed with a "b" following the year of publication.

> **Example:**
>
> (Foster, 1999a, 1999b).

Incorporating a Citation Into the Sentence

Up to this point, we have discussed how to cite to a source at the end of a sentence. Although it is an acceptable way to cite, citing after every sentence makes your writing choppy and less interesting to read. Moreover, it is important that you use variety in your writing, because it will help your paper flow more smoothly. Accordingly, one way to remedy the choppiness associated with citing at the end of every sentence is to incorporate citations into the beginning of a sentence.

UNIT 1

UNIT 2

UNIT 3

UNIT 4

UNIT 5

UNIT 6

UNIT 7

UNIT 8

UNIT 9

Examples:

a. According to Oakley (2001), boys commit more crimes than girls do.

b. Oakley (2001) found that boys commit more crimes than girls do.

General Rules

When you incorporate citations into the sentence, follow these rules:

1. Write the name(s) of the author(s) immediately followed by the publication year in parentheses. Do not write the name(s) of the author(s) at the beginning of the sentence and the year at the end.
2. Always include the year of publication in parentheses.
3. Do not include the title of the article in the text.
4. If you have incorporated the citation into the sentence, do not also write it at the end of the sentence.

Citing to One Author

When a source has one author, write the name as part of the text followed by the year of publication in parentheses.

Example:

According to Daniels (2009), DUI courts are successful in reducing recidivism.

Citing to the Same Source With One Author Within a Paragraph

When you incorporate a citation into the beginning of a sentence, you can make your writing flow more smoothly by using some variety in the way you start each sentence. In particular, instead of starting every sentence with the author's name—for example, "According to Daniels (2009) . . ."—you can use pronouns.

Example:

According to Williams (2010), inmates who participate in parenting classes learn to show compassion for others. She further stated that they become more self-confident about their parenting skills. Therefore, she concluded that more prisons should offer these classes at their facilities.

If you use pronouns, there are several rules you must follow:

1. Write the author's name in the sentence that precedes the first sentence in which you use a pronoun.
2. Use pronouns only when you cite to the same source in *the same paragraph*.
3. If you use a pronoun, use the correct gender for the author.

4. After you have cited to the author's name and year of publication, if you use his or her name again instead of a pronoun, do not include the year in subsequent citations in that paragraph.

Example:

According to Williams (2010), inmates who participate in parenting classes learn to show compassion for others. She further stated that they become more self-confident about their parenting skills. Therefore, Williams concluded that more prisons should offer these classes at their facilities.

5. If there is an intervening citation between your citations to Williams, you must write her name again, and you must include the year of publication in the citation.

Example:

According to Williams (2010), inmates who participate in parenting classes learn to show compassion for others. Wilson (2008) similarly found that the inmates become more self-confident about their parenting skills. Therefore, Williams (2010) concluded that more prisons should offer these classes at their facilities.

Citing to the Same Source With One Author in Subsequent Paragraphs

When you cite to a source in subsequent paragraphs, you must include the year again the first time you cite to the source in the new paragraph. For subsequent citations to that source within that paragraph, the same rules apply that were previously discussed.

Citing to Two Authors

When you cite to two authors in the text of your sentence, separate their names with the word *and,* and write the date of publication in parentheses.

Example:

According to Wilson and Black (2010), there are many benefits to prison education programs.

Citing to the Same Source With Two Authors Within a Paragraph

You can make your paper flow more smoothly by using the pronoun *they* when you cite to the same source with two authors.

Example:

Wilson and Black (2010) found that there are many benefits to prison education programs. In particular, they found that inmates who successfully complete a GED program are more likely to find employment when they are released. They found similar results for inmates who participate in vocational training classes.

UNIT 1

UNIT 2

UNIT 3

UNIT 4

UNIT 5

UNIT 6

UNIT 7

UNIT 8

UNIT 9

If you use pronouns, there are several rules you must follow:

1. Write the authors' names in the sentence that precedes the first sentence in which you use a pronoun.
2. Use pronouns only when you cite to the source by the same authors in *the same paragraph*.
3. After you have cited to the authors' names and year of publication, if you use their names again instead of a pronoun, do not include the year in subsequent citations in that paragraph.

Example:

Wilson and Black (2010) found that there are many benefits to prison education programs. In particular, they found that inmates who successfully complete a GED program are more likely to find employment when they are released. Wilson and Black found similar results for inmates who participate in vocational training classes.

4. If there is an intervening citation, include the authors' names and the year of publication in the next citation to that source.

Example:

Wilson and Black (2010) found that there are many benefits to prison education programs. According to Clover and Otis (2011), inmates who successfully complete a GED program are more likely to find employment when they are released. Wilson and Black (2010) found similar results for inmates who participate in vocational training classes.

Citing to the Same Source With Two Authors in Subsequent Paragraphs

When you cite to a source with the same two authors in subsequent paragraphs, you must include the year the first time you cite to the source. For subsequent citations to that source within that paragraph, the same rules that were previously discussed apply.

Citing to Three, Four, or Five Authors

When you cite to three, four, or five authors, separate the next-to-last author's and the last author's names with the word *and*. In addition, include a comma before the *and*. Then write the publication year in parentheses.

Example:

According to Oakley, Wilson, and Clover (2011), youth who are abused or neglected are more likely to commit crimes as adults.

The first time you cite to three, four, or five authors in your paper, you must include all of the names. Every subsequent time you cite to that source in your paper, write the citation as Oakley et al. (2011).

Citing to the Same Source With Three to Five Authors Within a Paragraph

After you have written the full citation in a paragraph, if you cite to Oakley et al. again in that paragraph and there are no intervening citations, you do not need to put the year of publication.

Example:

According to Oakley, Wilson, and Clover (2011), boys commit more crimes than girls do. However, Oakley et al. have found that most of the crimes they commit are misdemeanors.

When your source has three to five authors, you can also use the pronoun *they* to make your paper flow more smoothly. Follow the same rules discussed in the section about using *they* when citing to a source with the same two authors.

Example:

According to Oakley, Wilson, and Clover (2011), boys commit more crimes than girls do. However, they have found that most of the crimes they commit are misdemeanors.

If there is an intervening citation in a paragraph between two Oakley citations, you must put the first author's name followed by et al., and the publication year in parentheses in the next citation to that source.

Example:

According to Oakley, Wilson, and Clover (2011), boys commit more crimes than girls do. However, Otis (2011) found that most of the crimes they commit are misdemeanors. Further, Oakley et al. (2011) found that some boys do commit more serious crimes, such as assault and burglary.

Citing to the Same Source With Three to Five Authors in Subsequent Paragraphs

When you cite to the same source with three to five authors in subsequent paragraphs, you must include the year the first time you cite to the source. Remember, you do not need to write out all of the authors' names; write the citation as Oakley et al. (2011). For subsequent citations to that source within that paragraph, the same rules that were previously discussed apply.

Citing to Six or More Authors

When you cite to six or more authors in the text, write just the first author's last name followed by et al. and then the year of publication in parentheses.

Example:

According to Oakley et al. (2001), boys commit more crimes than girls do.

Citing to the Same Source With Six or More Authors Within a Paragraph

Once you have given the full citation, you do not need to include the year again in that same paragraph.

When your source has six or more authors, you can use the pronoun *they* to make your paper flow more smoothly. Follow the same rules discussed in the section about using *they* when citing to a source with the same two authors.

> **Example:**
>
> According to Wilson et al. (2011), boys commit more crimes than girls do. However, they have found that most of the crimes they commit are misdemeanors.

If there is an intervening citation in that paragraph, include the first author's name followed by et al. and the year of publication in parentheses in the second citation to that source.

> **Example:**
>
> According to Wilson et al. (2011), boys commit more crimes than girls do. However, Clover (2011) found that most of the crimes they commit are misdemeanors. Further, Wilson et al. (2011) found that some boys do commit more serious crimes, such as assault and burglary.

Citing to the Same Source With Six or More Authors in Subsequent Paragraphs

When you cite to those authors in subsequent paragraphs, include the year the first time you cite to that source. For subsequent citations to that source within that paragraph, the same rules apply that were previously discussed.

Combining In-Sentence Citations With Citations at the Ends of Sentences

When you write your paper, the best way to cite to your sources is by using a combination of in-sentence citations and citations at the ends of your sentences. By using variety in the way you cite, your paper will flow more smoothly and will be more interesting for the reader.

You can combine in-sentence citations with citations at the ends of sentences in different ways:

- Write some citations at the ends of your sentences and some as part of your text.

> **Example:**
>
> Researchers have found that boys commit more crimes than girls do (Oakley, Wilson, & Clover, 2001). Specifically, White and Johnson (2005) found that most of the crimes boys commit are misdemeanors. However, further research has revealed that some boys do commit more serious crimes, such as assault and burglary (Charles, Baldwin, & Cooper, 2008).

- Include string citations at the ends of some sentences, and incorporate citations into the beginnings of other sentences.

Example:

Researchers have found that boys commit more crimes than girls do (Adams, 2005; Oakley, Wilson, & Clover, 2001). Specifically, Oakley et al. found that most of the crimes boys commit are misdemeanors. However, further research has revealed that some boys do commit more serious crimes, such as assault and burglary (Adams, 2005).

Note! In this example, you do not need a year following "Oakley et al." because there are no intervening citations between it and the full citation. However, you do need to include a year for the Adams citation because you must always cite the year when you include a parenthetical citation at the end of the sentence.

Miscellaneous Rules for Citing in the Text

There are important miscellaneous rules to follow when you cite in the text.

Citing to a Secondary Source

Often an author will state a proposition in his or her article and then cite to the source from which he or she got that information.

Example:

You are reading an article written by Moore (2010). In it, Moore writes, "Youths from single-parent homes are more likely to become truant (Yardley, 2005)."

Here, Yardley is the primary source and Moore is the secondary source. When you want to use that information in your paper, locate and read Yardley's article and then cite to it in the text and on your reference list. However, it is not always possible to obtain an original source. If you are unable to locate the original source, cite to Moore on your reference list (because that is the source you actually read) and refer to Yardley's article in your text.

Example:

Yardley (2005) found youths from single-parent homes are more likely to become truant (as cited in Moore, 2010).

UNIT 1

UNIT 2

UNIT 3

UNIT 4

UNIT 5

UNIT 6

UNIT 7

UNIT 8

UNIT 9

Incorporating a Quote Into a Sentence

When you cite to a direct quote, include the name(s) of the author(s) and the page(s) from which you got the quote. There are two ways you can incorporate a quote into a sentence:

1. Write the name(s) of the author(s) and the year of publication as part of your text and then write the page number in parentheses after the quoted material. Place a period after the parentheses.

Example:

Oakley and Wilson (2008) found that "violent offenders are four times more likely to recidivate than nonviolent offenders" (p. 273).

2. Write the name(s) of the author(s) and the year of publication at the end of your sentence in parentheses. Include the page number in the parentheses. Place a period after the parentheses.

Example:

"Violent offenders are more likely to recidivate than nonviolent offenders" (Oakley & Wilson, 2001, p. 23).

See the following handouts at the end of this unit:

Citing in the Text in APA Style, Class Exercise #1: "Identifying Errors According to APA Style for Citing in the Text "

Citing in the Text in APA Style, Class Exercise #2: "Identifying Real-Life Errors According to APA Style for Citing in the Text"

Citing in the Text in APA Style, Handout #1: "Reference Guide"

Citing in the Text in APA Style, Handout #2: "APA Citing in the Text: A Real-Life Example"

NOTES

NOTES

UNIT
1

UNIT
2

UNIT
3

UNIT
4

UNIT
5

UNIT
6

UNIT
7

UNIT
8

UNIT
9

Citing in the Text in APA Style, Class Exercise #1

Identifying Errors According to APA Style for Citing in the Text, Part I

Read the following sentences and determine whether the underlined citations are correctly written according to APA style. If they are correct, write correct *above them. If they are incorrect, rewrite them so they are correct.*

1. There are many problems with boot camps <u>(J. Knowles, 2001).</u>

   _____

2. Juvenile drug use is increasing in the United States <u>(Oakley, Smith & Jones, 2008).</u>

   _____

3. <u>Miller</u> conducted a study that examined whether boys from single-parent homes are more likely to commit crimes than boys who are raised in homes with two parents <u>(2009).</u>

   _____

4. In his study, <u>Sternberg, (2010)</u> discussed how the social learning theory related to juvenile violence. In the study, Sternberg <u>(2010)</u> also briefly discussed the social control theory.

   _____

5. <u>Garcia & Smith (2007)</u> found that elder abuse is common in nursing homes. <u>Garcia et al.</u> also found that most of the abuse is committed by staff members.

   _____

6. <u>Basil, Williams, Clover, Daniels, Robertson, Edwards, and Smith (2009)</u> found that children who are abused are more likely to abuse their own children.

   _____

7. <u>Kelly, Stark, and Bert (2010)</u> found that 7% of teenagers drank alcohol and then drove. In another study, <u>Gamble and Clark</u> found that 85% of all automobile accidents involving teenagers also involved alcohol. Based on these studies, <u>Kelly, Stark, and Bert</u> concluded that programs must be developed to address the problem of teenagers who drink and drive.

   _____

8. One example of an alternative sanction for juveniles is boot camps <u>(Neal, 1999; Little & Johns, 2002).</u>

   _____

9. Punishment is the foundation of the deterrence theory <u>(Smith, 2009; Lucker and Osti, 2001; Smith & Travis, 2000).</u>

   _____

UNIT 1

UNIT 2

UNIT 3

UNIT 4

UNIT 5

UNIT 6

UNIT 7

UNIT 8

UNIT 9

Citing in the Text in APA Style, Class Exercise #1

Identifying Errors According to APA Style for Citing in the Text, Part II

Rewrite the following sentences using correct citations:

1. In an article published in 2008, Smith found that the number of juveniles sentenced to boot camps has decreased.

   _____

   _____

   _____

2. Boys are more likely to skip school than girls are (Smith, 2001). In addition, boys are more likely to stay out past midnight (Smith, 2001). Finally, boys are more likely to smoke cigarettes than girls are (Smith, 2001).

   _____

   _____

   _____

UNIT 1

UNIT 2

UNIT 3

UNIT 4

UNIT 5

UNIT 6

UNIT 7

UNIT 8

UNIT 9

Citing in the Text in APA Style, Class Exercise #2

Identifying Real-Life Errors According to APA Style for Citing in the Text

Read the following paragraphs and determine whether the citations are correctly written. If a citation is correct as written, write correct *above it. If it is incorrect, revise it to be correct. If a citation is included where it should not be included, make a mark through it. Finally, if a citation is missing, note where it should be included.*

The following material has been modified from Ferree, C. W. (2006). *DUI recidivism and attorney type: Is there a connection?* (Unpublished master's thesis). School of Criminal Justice, University of Baltimore, Baltimore, MD.

1. Each year, millions of indigent defendants rely on publicly appointed counsel for legal representation, and each year, enormous sums of money are spent on their defense. For example, according to DeFrances & Litras (2000), in 1999, public counsel represented indigent offenders in approximately 4.2 million cases (DeFrances & Litras). Moreover, that same year taxpayers spent an estimated $1.2 billion on indigent legal defense in the 100 most populous counties in the United States.

2. In order to decrease recidivism rates, courts have imposed various sanctions on offenders (James, Walker, Smith, Jones, Tyler, & Baker)(2001). Historically, individuals who were convicted of alcohol-impaired driving were punished through incarceration (Lucker and Osti, 1997). They stated that this approach was based on a deterrence theory under which policymakers believed that fewer individuals would drink and drive if they faced swift, certain, and severe punishments. In accordance with this policy, as of 1988, 42 states had adopted laws mandating incarceration for repeat DWI offenders (Martin, Annan, & Forst, 1993; Lucker & Osti, 1997). In addition, 14 states had passed laws mandating jail sentences for first-time DWI offenders (Lucker et al., 1997).

The following three paragraphs go together as an excerpt from a paper. For this section, if a citation is correct as written, write correct *above it. If it is incorrect, rewrite it correctly. If a citation is included where it should not be included, make a mark through it. Finally, if a citation is missing, note where it should be included.*

3. According to Neinstadt, Zatz, and Epperlein (1998), of the millions of defendants who use appointed counsel, many defendants share similar characteristics. In particular, indigent defendants are likely to be either unemployed or employed part time, and are typically from a lower socioeconomic class (Champion, 1989; Nienstadt, Zatz, & Epperlein, 1998; Sterling, 1983; Baker and Jones, 2000). They are also likely to have prior criminal records, be young, and have low levels of education (Sterling, 1983).

UNIT 2

UNIT 3

UNIT 4

UNIT 5

UNIT 6

UNIT 7

UNIT 8

UNIT 9

In addition to the similarities among the personal characteristics of indigent defendants, studies have found similarities in the crimes with which they are charged. According to Neinstadt, Zatz, & Epperlein, in state courts, indigent defendants who are represented by public attorneys are more likely to be charged with serious crimes (e.g., violent crimes, property offenses, and drug crimes) (1998). In contrast, Harlow (2000) found that offenders who are represented by private attorneys are more likely to be charged with less serious crimes, such as public order offenses (including driving offenses) (Harlow, 2000).

Studies have also shown that offenders who hire private attorneys share certain characteristics (Baker & Jones, 2000). In particular, in 1983, Sterling found that defendants who are represented by private counsel are more likely to be charged with narcotics offenses than with more serious felonies such as robbery or burglary. It was also found that those offenders were less likely to have a prior record, or to have been out on bail, probation, or parole at the time of the commission of the offense.

UNIT 1

UNIT 2

UNIT 3

UNIT 4

UNIT 5

UNIT 6

UNIT 7

UNIT 8

UNIT 9

Citing in the Text in APA Style, Handout #1

Reference Guide
End-of-Sentence Citations

One author: (Smith, 2001).
Two authors: (Smith & Jones, 2001).
Three to five authors:
> First time: (Oakley, Wilson, & Jones, 2001).
> Every subsequent time: (Oakley et al., 2001).

Six or more authors (every time): (Wilson et al., 2001).
Government agency as author:
> First time: (National Institute of Health [NIH], 2001).
> Every subsequent time: (NIH, 2001).

Several sources, same information (alphabetical by first author's last name):
> (Oakley & Smith, 2001; Smith & Jones, 2003).
> (Oakley, 2003; Smith, 2001; Wilson, Edwards, & Jones, 2000).
> (Oakley, 2003; Smith, 2001; Smith & Brown, 2000).

Same author, different years: (Smith, 1997, 2001).
Same author in same year (in reference list):
> Smith, J. C. (1999a). An Analysis of ... In the text—(Smith, 1999a)
> Smith, J. C. (1999b). Evaluation of ... In the text—(Smith, 1999b)

In-Text Citations

One author: According to Smith (1999) ... (NOT, "In 1999, Smith found ....")
Two authors: In contrast, Smith and Jones (2003) argued that ... They also found ...
Three to five authors:
> First time: According to Smith, Jones, Walker, and Gable (2000) ...
> Every subsequent time: Furthermore, Smith et al. (2000) noted ...

Six or more authors (every time): According to Walker et al. (2003) ...

> **Rule:** Once you introduce another source, you must give a full citation:
> Source A → Source B → Source A.

Example:

According to Oakley, Wilson, and Clover (2011), boys commit more crimes than girls do. However, Otis (2011) found that most of the crimes they commit are misdemeanors. Further, Oakley et al. (2011) found that some boys do commit more serious crimes, such as assault and burglary.

UNIT 1

UNIT 2

UNIT 3

UNIT 4

UNIT 5

UNIT 6

UNIT 7

UNIT 8

UNIT 9

Citing in the Text in APA Style, Handout #2

APA Citing in the Text: A Real-Life Example

The following material has been modified from Ferree, C. W. (2006). *DUI recidivism and attorney type: Is there a connection?* (Unpublished master's thesis). School of Criminal Justice, University of Baltimore, Baltimore, MD.

Over the past several decades, alcohol-impaired driving has become a very serious problem in the United States. According to the Centers for Disease Control and Prevention (CDCP, 2004), in 2002, 1.5 million people were arrested for driving while impaired. Moreover, studies have shown that alcohol-related automobile accidents are a leading cause of death and physical injury (Marzano, 2004; Meyer & Gray, 1997; National Highway Traffic Safety Administration [NHTSA], 2005). It is estimated that each year 120 million people drink and drive, although the exact number is difficult to determine because many people drink and drive but are never caught (Marzano, 2004; Voas & Fisher, 2001). In fact, Breer, Schwartz, Schillo, and Savage (2003) estimated that individuals might drive 1,000 times while under the influence of alcohol before they are stopped by the police.

Although driving under the influence of alcohol is one of the greatest public safety concerns in the United States, Voas and Fisher (2001) report that the number of alcohol-related traffic fatalities has decreased by approximately 20% over the past 20 years. Similarly, Breer et al. (2003) found that from 1993 through 2003, the number of individuals who drove while intoxicated also decreased. Despite this decrease, alcohol-related accidents occur across the United States at a staggering rate. According to the CDCP (2004), someone is killed in an alcohol-related automobile accident every 31 minutes, and someone is injured in such an accident every 2 minutes. The CDCP also found that every year, 40% of all traffic-related deaths are alcohol-related. In total, in 2003 and 2004, approximately 17,000 people were killed as a result of alcohol-related driving accidents (NHTSA, 2005).

UNIT 1

UNIT 2

UNIT 3

UNIT 4

UNIT 5

UNIT 6

UNIT 7

UNIT 8

UNIT 9

Preparing for the Job Market

UNIT SUMMARY

Learning Objectives

At the end of this unit, students will be able to write a chronological résumé.

Writing a Résumé: An Overview

The first step in the hiring process is to provide a prospective employer with a résumé. Because the résumé will be the first impression you make, you must be sure that it is well organized, error-free, and succinct. Employers typically receive dozens, if not hundreds, of résumés for a job opening and have only a limited amount of time to spend reviewing each one. Thus, you should make it as concise as possible, highlighting your accomplishments in an easy-to-read and professional format.

When you write your résumé, it is imperative that you use correct spelling and proper grammar. Employers who receive résumés with spelling and grammar errors typically assume that you will make similar mistakes at work, and they will reject your application.

 Note!   Many students erroneously believe that the purpose of a résumé is to get a job. It is not. The purpose of a résumé is to get an interview, so that you can discuss in person, with your prospective employer, your qualifications for the job.

There are many different types of résumés. In this unit, you will learn how to write a chronological résumé.

Example:

State the definition of a chronological résumé.

_____

_____

In your résumé, you will organize your information into six categories. However, if you do not have experience in one or more of the categories, you will exclude it.

Example:

List the six categories included in a résumé.

_____

_____

_____

Writing a Résumé: The Basic Rules About Appearance

One of the most important things to remember when you write a résumé is that it must have a professional appearance. Accordingly, follow these basic rules:

   1. Use a traditional font that is easy to read.

Example:

Write examples of two types of fonts you can use to write your résumé.

_____

2. With the exception of the font you use for your name, use the same font size throughout your résumé. Make your name one font size larger than the font size you select for the rest of the résumé.

Example:

What size font should you use when you write your résumé?

_____

3. Use heavy bond in a neutral-color paper such as linen or off-white.

4. Limit your résumé to one or two pages. At the interview, you will be able to discuss in detail your qualifications and experiences.

Writing a Résumé: The Basic Rules About Content

The content of your résumé must be well written and presented in a succinct, professional manner. To ensure you do this, follow these guidelines:

- Include only the information that you feel makes you a strong candidate for the position and that you feel would impress an employer. For example, include information about a job in which you were given supervisory duties, but do not include a job if you only held it for a few weeks or if you left it on unfavorable terms.
- When you write your résumé, use spell check. However, remember that it will not catch all errors, such as when you use the wrong homophone.
- Tailor your résumé so that it is geared toward the position for which you are applying. For example, if you have experience in both corrections and store security, and you are applying for a corrections position, you should tailor your résumé toward your experience in corrections.

In addition to the rules governing what you should include in your résumé, there are also rules about what you should _not_ include.

Example:

Write three examples of information you should not include in a résumé.

_____

_____

_____

Writing a Résumé: Getting Started

When you write a basic chronological résumé, you will present your information in six sections. Therefore, the first thing you should do is make a list of the sections you will include. List the sections in the same order in which they will appear in your résumé. Once you have made your list, write all of the information you want to include in each of those sections.

Education Information

For this section, list the academic institutions you attended after graduating from high school. These include trade schools, community colleges, and 4-year colleges or universities that you either attended or from which you graduated. For each school, write the name, location (city and state), and year(s) you attended.

If you attended a school for less than a year, write the months and year that you attended the school. If you are in the process of receiving a degree from a school, write your expected date of graduation. If you have received a degree from any of the schools you listed, write the name of the degree and your discipline of study (e.g., Bachelor of Science in Criminology), and the date on which you received it. If you graduated with honors or with any other distinction (e.g., cum laude), include that information as well. Do not, however, include your GPA. The employer can verify that information from your transcripts if it is pertinent.

Employment Information

For this section, list all of the jobs you have held since high school graduation. Include all part-time and full-time jobs, as well as internships (paid and unpaid) that you have held. If you have been out of high school for more than 10 years, list only the jobs you have held within the past 10 years. For each job, write the name of the business, the city and state where it is located, the dates you were employed, your job title, and your job duties. If you did not have an official job title, create one that best describes your position.

Certifications and Technological Skills

For this section, list the certifications or special training you have received, and technological (e.g., computer-related) skills you have that are related to the job you are seeking. In this section, you should also list any foreign languages in which you are proficient.

When you list your computer-related skills, highlight only those computer programs in which you are proficient that would require additional training to master, such as statistical packages (e.g., SPSS) or specialized databases (e.g., LexisNexis, CompSTAT). Do not include programs that most employers assume candidates would be proficient in (e.g., Word, PowerPoint).

Academic and Professional Honors

For this section, list all of the academic and professional honors you have received since you graduated from high school.

Example:

Write some examples of honors you could include in your résumé.

_____

_____

UNIT
1

UNIT
2

UNIT
3

UNIT
4

UNIT
5

UNIT
6

UNIT
7

UNIT
8

UNIT
9

Community and Professional Engagement

For this final section, include all activities that reflect your contribution to your community and profession.

Example:

Write some examples of community and professional engagement activities you could include in your résumé.

_____

_____

Writing a Résumé: The Details

You must follow certain rules to ensure that your résumé is presented professionally, because its appearance is the first impression you will make on a prospective employer. To help you better understand how to present a professional-looking résumé, this section discusses the sample résumé located at the end of this unit.

See **Preparing For the Job Market, Handout #1: "Sample Résumé"** at the end of this unit.

The Appearance

There are certain things you should note about the overall appearance of the sample résumé:

- The name is typed in bold, in a font size that is larger than the font size used in the rest of the résumé.
- Each heading is capitalized and in bold (e.g., objective, education).
- The dates relating to education and work experience are set apart from the written content.
- All of the information included within each category is presented in succinct phrases. There are no periods at the end of the phrases.
- Each job duty, certification, technological skill, honor, and community and professional engagement activity is set in a bulleted list.
- The résumé is double-spaced between the categories and between each entry in a category. However, each entry is single-spaced.
- All of the information for the Related Experience section fits onto the first page. However, you may find that information in this section (or another section) must be split over two pages. If this is the case, do not break entries; insert a page break so the entire entry appears on the second page.

Example:

On the sample résumé, if all of the information for the Best Buy job had not fit onto the first page, Jane Doe would have inserted a page break directly after the previous job description—Nordstrom—so that the entire Best Buy section would appear on the second page.

The Specific Sections

As well as having a professional appearance, your résumé must include specific information. Before listing the categories discussed above, your résumé will begin with your personal Identifying Information—also called Contact Information—and your Objective. Again, to better understand what each section must include, refer to the sample résumé handout at the end of this unit.

Identifying Information

There are several things you should note about the identifying information in the sample résumé:

- The identifying information is centered at the top of the résumé.
- The email address is professional. If you do not have a professional email address, you should create one.
- There is one telephone number. However, if you can be easily reached at several numbers, include all of them on your résumé.

The Objective

The objective sets the tone for your résumé. Its purpose is to allow you to briefly tell the employer what you will bring to the job. You should make it job specific and succinct.

When you write your objective, think about the position you are seeking and any special skills or interests you have that would strengthen your qualifications for that position. Often, it is helpful to refer back to the position advertisement when you write your objective. Specifically, when you describe yourself in your objective, you can use the adjectives that have been specified as necessary or desirable for the potential candidate to have.

 In the sample résumé, the objective is very short and does not include full sentences. Moreover, it briefly highlights the applicant's attributes (e.g., current education level, position sought, length of related work experience, and relevant certification).

Education

In this section, you should include all of your post-secondary education information starting with the most recent. For each degree you have earned, in italics, write the title of the degree and discipline in which you received it, followed by a comma and the year you received it, also in italics. Below that, in regular font, write the full name of the school you attended and the city and state where it is located. When you write the state's abbreviation, write it in capital letters and do not include periods. If you graduated with honors from a school, include that information.

Example:

If you earned a degree:

EDUCATION

Bachelor of Science in Criminal Justice, 2005
Roanoke College, Roanoke, VA
- Graduated with honors

If you are in the process of earning a degree, write the title of the degree and discipline in which you will receive it in italics, followed by a comma. Below that, in regular font, write the full name of the school you attend and the city and state where it is located. To the right, write the date you expect to receive your diploma so that it is on the same line as the degree you will earn and insert a right margin tab so that it is flush with the right margin.

Example:

If you are in the process of earning a degree:

EDUCATION

Bachelor of Science in Criminology,   Expected in May 2013
University of Baltimore, Baltimore, MD

If you attended a school but did not graduate from it, list the name of the school, its location, and the years you attended. If you attended it for less than 1 year, include the months and year you attended. If you attended more than one school from which you did not graduate, list each one.

Example:

If you attended an institution but do not anticipate graduating from that school:

EDUCATION

Bucknell University, Lewisburg, PA   January 2011–June 2011

Related Experience

In this category, starting with your most recent job, list your work experience in fields that are related to the job you are seeking.

Note! If you have not had any experience in a related field, you will not include a Related Experience section. Instead, you will list all of your work experience in your Employment History, which will be discussed in the next section.

Your Related Experience includes all full-time work, part-time work, and internships (paid or unpaid). For each job you held (or currently hold), list the name of the company in capital letters. Following that, in lowercase letters, write the name of the city and, in capital letters, write the abbreviation for the state where you work(ed). To the right, insert a right margin tab and write the dates of your employment so they are flush with the right margin.

Beneath the name of the company, in bold font, write your job title. If you did not have an official title, create one that best describes your position. Beneath that, using bullets, write *short* phrases stating your most important job duties. Always list at least two duties, but do not list more than five. In addition, do not write lengthy explanations for what you did, and do not write *etc*. You can explain your job duties in more detail to the employer during your interview.

UNIT 1
UNIT 2
UNIT 3
UNIT 4
UNIT 5
UNIT 6
UNIT 7
UNIT 8
UNIT 9

When you state your job duties at your current job, write your verbs in the present tense (e.g., supervise 25 employees). When you describe the job duties you performed in a past job, write your verbs in the past tense (e.g., prepared theft reports).

When you state your job duties, use forceful verbs and write objectively; do not sing your own praises. If you wrote well or performed your job well, that will be reflected in your references.

Example:

Do not say: Was responsible for preparing well-written and thorough theft reports for management.
Instead, simply say: Prepared theft reports for management.

Employment History

Present your past and current employment information in the same way that you presented your Related Experience information; however, in this section, include all work experience you have had that is *not* related to the job you are seeking. If all of your work experience is related to the job you are seeking, you will not include this section.

Again, you should include all full-time work, part-time work, and internships (paid or unpaid), beginning with the most recent.

 Note!   In the sample résumé, all of Jane Doe's most recent jobs were related to the one she was seeking, and the older ones were not. However, it is possible that your work experience may go back and forth between the two categories so that the correct chronological order is split between the two. This is acceptable.

Certifications and Technological Skills

In this section, list your certifications and technological (computer) skills in regular font and in short, bulleted phrases. Also list any foreign languages in which you are proficient. Remember, list only the computer skills that required additional training; do not list basic computer skills.

Academic and Professional Honors

In this section, include all of the academic and professional honors you have received since you graduated from high school. For each honor you list, first write the honor in regular font. Then write the name of the institution from which you received that honor. Beneath that, also in regular font and using bullets, list the relevant dates, starting with the most recent one.

Example:

ACADEMIC AND PROFESSIONAL HONORS

Dean's List—University of Baltimore
• Spring 2011, Fall 2010

Community and Professional Engagement

In this section, list the activities that indicate your contribution to your community and profession. The format of these entries is similar to that used in the academic and professional honors section. In addition, if you held (or hold) a position in an organization, write the name of the position in italics followed by a comma and then the relevant dates (in regular font).

References

The last thing you should include on your résumé is the brief statement: References available upon request. Write it in regular font, without bold or italics, at the bottom of the page, as shown on the sample résumé. Do not include a list of references on your résumé; you can provide that to the employer in a separate document if he or she asks for it.

There are certain guidelines you should follow when choosing your references.

- Make sure a prospective reference is willing to act as a reference for you before you give his or her name to your interviewer. Not only is it polite and professional to do so, but it will also allow you to determine whether that individual's referral will be favorable. If the individual hesitates, or seems uncomfortable, you should choose a different reference.
- Prepare an updated copy of your résumé for your reference so that he or she can review your accomplishments and speak more knowledgably about you to the employer.
- If possible, choose as a reference someone who is (or has been) employed in the field in which you are seeking a job.
- It is acceptable to ask a professor to act as a reference. However, choose from among only those professors with whom you have had a class within the past 2 years; longer than that may indicate to the prospective employer that you are unable to find a more recent referral. In addition, make sure you choose a professor in whose class you did well!

> If your instructor will have you complete the Writing a Résumé Assignment, see **"Preparing for the Job Market, Handout #2: Writing a Résumé Assignment"** at the end of this unit.

UNIT 1
UNIT 2
UNIT 3
UNIT 4
UNIT 5
UNIT 6
UNIT 7
UNIT 8
UNIT 9

NOTES

NOTES

UNIT 1

UNIT 2

UNIT 3

UNIT 4

UNIT 5

UNIT 6

UNIT 7

UNIT 8

UNIT 9

Preparing for the Job Market, Handout #1

235

UNIT
1

UNIT
2

UNIT
3

UNIT
4

UNIT
5

UNIT
6

UNIT
7

UNIT
8

UNIT
9

Preparing for the Job Market, Handout #1

Sample Résumé

Jane L. Doe
236 Main Street
Baltimore, MD 21209
(410) 333-5609
jane.doe@ubalt.edu

OBJECTIVE

Highly motivated and dependable bachelor's candidate seeking full-time position in federal law enforcement. Offers more than 4 years of security and management experience in the private sector and is certified as an Emergency Response Coordinator by the American Red Cross.

EDUCATION

Bachelor of Science in Criminology, Expected in May 2013
University of Baltimore, Baltimore, MD

RELATED EXPERIENCE

NORDSTROM—Towson, MD April 2010–present
Manager of Loss Prevention Operations
- Supervise 25 employees
- Hire and train all new employees
- Facilitate and consult in the development and implementation of new security protocols with corporate headquarters
- Oversee all investigations

NORDSTROM—Towson, MD March 2009–April 2010
Loss Prevention Specialist
- Prepared theft reports for management
- Assisted manager with employee training and loss prevention
- Created new security protocol for prevention of employee theft

BEST BUY—Towson, MD February 2007–March 2009
Loss Prevention Specialist
- Assisted manager in investigating in-store thefts
- Prepared loss reports

EMPLOYMENT HISTORY

REI—Timonium, MD   June 2006–January 2007
Sales Representative
- Assisted customers
- Organized and catalogued inventory
- Assisted manager in training new employees

PIZZA HUT—Towson, MD   October 2005–June 2006
Head Hostess
- Supervised waitstaff
- Trained all new front-staff employees
- Assisted manager with payroll

CERTIFICATIONS AND TECHNOLOGICAL SKILLS
- Certified as Emergency Response Coordinator through Red Cross of America, June 2009
- Proficient in SPSS and LexisNexis

ACADEMIC AND PROFESSIONAL HONORS

Dean's List—University of Baltimore
- Spring 2011, Fall 2010

COMMUNITY AND PROFESSIONAL ENGAGEMENT

Red Cross of America
- *Volunteer*, 2009 to present

Academy of Criminal Justice Sciences
- *Member*, Fall 2010 to present

Criminal Justice Student Association
University of Baltimore, Baltimore, MD
- *Member*, Fall 2010 to present
- *Officer (Secretary)*, Spring 2011 to present

References available upon request

UNIT 1
UNIT 2
UNIT 3
UNIT 4
UNIT 5
UNIT 6
UNIT 7
UNIT 8
UNIT 9

UNIT 1

UNIT 2

UNIT 3

UNIT 4

UNIT 5

UNIT 6

UNIT 7

UNIT 8

UNIT 9

Preparing for the Job Market, Handout #2

Writing a Résumé Assignment

Students must write a chronological résumé following the format taught in class. Résumés must be printed on résumé-quality paper and should be no more than two pages long.

References

American Psychological Association. (2010). *Publication manual of the American Psychological Association* (6th ed.). Washington, DC: Author.

Espinel, V. (2011, March 15). Concrete steps Congress can take to protect America's intellectual property (Web log post). Retrieved from http://www.whitehouse.gov/blog/2011/03/15/concrete-steps-congress-can-take-protect-americas-intellectual-property

Etter, G. W., Sr., & Birzer, M. L. (2007). Domestic violence abusers: A descriptive study of the characteristics of defenders in protection from abuse orders in Sedgwick County, Kansas. *Journal of Family Violence, 22,* 113–119. doi: 10.1007/s10896-006-9047-x

Ferree, C. W. (2006). *DUI recidivism and attorney type: Is there a connection?* (Unpublished master's thesis). University of Baltimore, Baltimore, MD.

Finkelhor, D., & Ormrod, R. (2001). *Crimes against children by babysitters* (Juvenile Justice Bulletin). Washington, DC: U.S. Department of Justice.

McCabe, D. L., Trevino, L. K., & Butterfield, K. D. (2001). Cheating in academic institutions: A decade of research. *Ethics & Behavior, 11*, 219–232.

National Institute of Justice. (2006). *Drug courts: The second decade* (Special Report). Washington, DC: Author.

Office for Victims of Crime. (2003). *Community outreach through police in schools* (OVC Bulletin). Washington, DC: U.S. Department of Justice.

Rape, Abuse, and Incest National Network. (2009). *Reporting rates.* Retrieved from http://www.rainn.org/get-information/statistics/reporting-rates

Rennison, C. M. (2002). *Rape and sexual assault: Reporting to police and medical attention.* Washington, DC: Bureau of Justice Statistics.

U.S. Census Bureau, Population Division. (1995). *Sixty-five plus in the United States* (Statistical Brief). Retrieved from http://www.census.gov/population/socdemo/statbriefs/agebrief.html

Widom, C. S. (1995). *Victims of childhood sexual abuse—later criminal consequences* (Research in Brief). Washington, DC: National Institute of Justice.